THE GLOBAL COPPER INDUSTRY

Problems and Prospects

RAYMOND F. MIKESELL

CROOM HELM
London • New York • Sydney

Croom Helm Ltd, Provident House, Burrell Row,
Beckenham, Kent, BR3 1AT
Croom Helm Australia, 44-50 Waterloo Road,
North Ryde, 2113, New South Wales

Published in the USA by
Croom Helm
in association with Methuen, Inc.
29 West 35th Street
New York, NY 10001

British Library Cataloguing in Publication Data

Mikesell, Raymond F.
The global copper industry: problems and prospects.
1. Copper industry and trade
I. Title
338.4'76693 HD9539.C6

ISBN 0-7099-3508-0

Library of Congress Cataloging-in-Publication Data

Mikesell, Raymond Frech.
The global copper industry: problems and prospects/Raymond F. Mikesell.
p. cm. — (Croom Helm commodity series)
Bibliography: p.
Includes index.
ISBN 0-7099-3508-0
1. Copper industry and trade. I. Title. II. Series
HD9539.C6M53 1988
338.2'743 — dc 19 87-23073
CIP

Printed and bound in Great Britain
by Billing & Sons Limited, Worcester.

60 0327186 5
TELEPEN

THE GLOBAL COPPER INDUSTRY:
Problems and Prospects

CROOM HELM COMMODITY SERIES
Edited by Fiona Gordon-Ashworth, Bank of England

URANIUM: A STRATEGIC SOURCE OF ENERGY
Marian Radetzki, Institute for International Economic Studies, Stockholm

TIN: ITS PRODUCTION AND MARKETING
William Robertson, University of Liverpool

INTERNATIONAL COMMODITY CONTROL: A Contemporary History and Appraisal
Fiona Gordon-Ashworth, Bank of England

COMMODITY MODELS FOR FORECASTING AND POLICY ANALYSIS
Walter C. Labys and Peter K. Pollak

THE MODERN PLANTATION IN THE THIRD WORLD
Edgar Graham with Ingrid Floering

THE INTERNATIONAL GRAIN TRADE: PROBLEMS AND PROSPECTS
Nick Butler

THE POLITICAL ECONOMY OF NATURAL GAS
Ferdinand Banks

COMMODITY POLICIES: PROBLEMS AND PROSPECTS
A.I. MacBean and D.T. Nguyen

The series is edited by Fiona Gordon-Ashworth, formerly of the University of Southampton, who now works at the Bank of England. (The views expressed in this book are not to be taken as those of the Bank of England.)

TABLE OF CONTENTS

AUTHOR'S PREFACE

Since the publication of my book, The World Copper Industry, in 1979, dramatic changes have taken place in the structure of the copper industry. Low copper prices and nearly flat demand have led to substantial shifts in the market shares of producing countries. U.S. production, which was the largest in the world in 1981, has fallen by one-third and some of the leading U.S. mining firms have withdrawn from the copper industry. During the 1980s growth in copper output has been confined to developing countries, led by Chile which has been responsible for the bulk of the increase in new capacity in the market economies. This book analyzes these developments and, on the basis of projections of copper demand and capacity, forecasts the likely end of the long period of current overcapacity and low prices.

Shortly after completing a first draft of my manuscript, I was privileged to see a draft of the World Bank staff study The World Copper Industry: Its Changing Structure and Future Prospects (1986) by Kenji Takeuchi, John E. Strongman, Shunichi Maeda and Suan Tan. Those portions of my book dealing with the future of the copper industry have greatly benefited from capacity and other projections in their study.

I would like to express my appreciation to Dr. Kenji Takeuchi of the World Bank and Dr. John W. Whitney of Whitney & Whitney Mining Consultants for their helpful comments on the manuscript. I also want to thank my secretary, Letty Fotta, for her painstaking work in preparing the manuscript, and my wife, Irene, without whose reference and editorial services none of my books would be written.

ABBREVIATIONS

AAC	Anglo-America Corporation of South Africa
AMAX	American Metals Climax
ARCO	Atlantic Richfield Company
ASARCO	American Smelting and Refining
ARCO	Atlantic Richfield Company
BCL	Bougainville Copper Ltd.
BCL	Bamangwato Copper Ltd.
BHP	Broken Hill Proprietary
BOM	U.S. Bureau of Mines
CCR	continuous cast rod
CIPEC	Intergovernmental Council of Copper Exporting Countries
CODELCO	Corporacion Nacional del Cobre de Chile
COMEX	New York Commodity Exchange
CRU	Commodities Research Unit
CRA	Charles River Associates
CVRD	Cia. Vale do Rio Doce
EPA	U.S. Environmental Protection Agency
EC	Economic Community
FDI	foreign direct investment
FTC	U.S. Federal Trade Commission
GATT	General Agreement on Tariffs and Trade
GDP	gross domestic product
Gecamines	Generale des Carrieres et des Mines
GNP	gross national product
IADB	Inter-American Development Bank
ICSID	International Centre for Settlement of Investment Disputes
IDA	International Development Association
IFC	International Finance Corporation
IMD	Integration of Multi-Disciplinary Data Sets
IMF	International Monetary Fund
INCO	International Nickel Company
IPC	Integrated Program for Commodities
IRR	internal rate of return
ISM	in-situ solution mining
ITA	International Tin Association
ITC	U.S. International Trade Commission
LANDSAT	multi-spectral satellite scanning
LME	London Metal Exchange
MNC	multinational corporation
mt	metric ton
mtpy	metric ton per year
NIEO	New International Economic Order

ABBREVIATIONS (cont.)

NPV	net present value
OBC	overland belt conveyors
OECD	Organization for Economic Cooperation and Development
OPEC	Organization of Petroleum Exporting Countries
OPIC	U.S. Overseas Private Investment Corporation
OPMC	on-line production monitoring and control
PNG	Papua New Guinea
R&D	research and development
ROR	rate of return
RRT	resource rent tax
RTZ	Rio-Tinto Zinc
SME	state mining enterprise
SOHIO	Standard Oil Company of Ohio
SO_2	sulfur dioxide
SMTF	Societe Mineriere de Tenke-Fungurume
SPCC	Southern Peru Copper Corporation
st	short tons
UK	United Kingdom
UN	United Nations
UNCTAD	United Nations Conference on Trade and Development
U.S.	United States
USGPO	United States Government Printing Office
U.S.S.R.	Union of Soviet Socialist Republics
ZCCM	Zambian Consolidated Copper Mines

DEFINITION OF COUNTRY AREAS

Industrial Countries

Canada, United States, Belgium-Luxembourg, Denmark, France, Federal Republic of Germany, Italy, Netherlands, United Kingdom, Austria, Norway, Sweden, Switzerland and Japan

Developed Countries

North America: Canada and the United States

European Economic Community: Belgium/ Luxembourg, Denmark, France, Federal Republic of Germany, Ireland, Italy, Netherlands, Spain, United Kingdom, Greece and Portugal

Other Western Europe: Austria, Finland, Norway, Sweden and Switzerland

Africa: Republic of South Africa

Asia and Oceania: Japan, Australia and New Zealand

Developing Countries

Southern Europe: Cyprus, Yugoslavia, and Turkey

Middle East: Egypt, Israel, Jordan, Lebanon, Saudi Arabia, and Syria

Africa: all African countries except South Africa

Latin America: Mexico, Central America, South America and Caribbean countries (excluding Cuba)

Asia and the Pacific Islands: (excluding Japan and Communist countries)

Communist or Centrally Planned Economies

Eastern Europe: Albania, Bulgaria, Czechoslovakia, German Democratic Republic, Hungary, Poland, Romania and USSR

Asia: China, Cambodia, North Korea, Laos, Mongolia, and Viet Nam

Latin America: Cuba

TABLES AND FIGURES

Chapter 1

SURVEY OF THE WORLD COPPER INDUSTRY

BRIEF HISTORY

Man has used copper for making weapons and tools as early as the fifth millenium BC. Copper metallurgy (smelting ore) was introduced sometime in the fourth millenium BC, followed by the discovery of bronze (a copper/tin alloy) early in the third millenium BC.[1] Bronze became the dominant metal for production of weapons and tools in the West until replaced by steel in the first millenium AD.[2] Some of the ancient copper mines--those in the Timna Valley in Israel (dating from 4000 BC), the Cyprus Mines (dating from 3000 BC) which supplied the Phoenicians, Greeks and Romans, and the Rio Tinto mines in Southern Spain (dating from 1500 BC)--have been rediscovered and are operating today.[3]

During the Middle Ages and the early modern period, Continental Europe was the center of the world copper industry. The munitions industry was the principal consumer for the production of brass cannon and other military items. By the end of the eighteenth century Britain was the world's largest copper producer, with most of the production in the Cornish and Devon areas. (These ores were exploited by the Phoenicians as early as 1500 BC and later developed by the Romans.) When the demand for copper exceeded Britain's mine capacity, it began importing ore from various parts of the world for smelting. British advances in smelting technology gave that country a near monopoly on copper output, which continued to the middle of the nineteenth century. Britain's position in the industry declined rapidly in the second half of that century with the growth of output in Chile and later in the U.S., Canada and Mexico. Meanwhile, world demand for copper began to double every few years due to the growth in the use of electric power for lighting and communications. World output, which was estimated at 100,000 metric tons (mt) in 1860, reached 1 million mt by 1912; 2 million mt by 1929; and nearly 8 million mt by the mid-1970s.

The rapid growth in world demand for copper was accompanied by a series of important discoveries of deposits in North and South America, Africa and later in the Far East. The first large deposits exploited in the U.S. were in the Lake Superior region of northern Michigan, an area where American Indians had been recovering native copper for centuries.[4] For a time after the Civil War, northern Michigan was the largest U.S. source of copper, but later in the century its output was exceeded by mines discovered at Butte, Montana. From 1850 to 1880 Chile was the world's largest copper exporter. A number of relatively small mines had been producing copper in Chile since the early part of the nineteenth century, but output was small compared to what it became in the twentieth century.[5]

Prior to development of technology for large-scale mining and processing of low-grade deposits, mining was limited to the extraction of rich underground veins containing 6 to 10 percent or more copper. Some of the early mines were able to operate profitably without smelters in isolated areas many hundreds of miles from a railroad because they produced ores containing 10 to 20 percent copper. The accessible high-grade ores were rapidly depleted, but far more mineable copper was available with grades under 2 percent. In 1905 an American engineer, Daniel C. Jackling, introduced mass production in an open pit mine which contained only 2 percent copper. About the same time, there were improvements in the concentrating process. The new mining and processing technologies made possible the creation of a number of large open pit mines in several Western states, including Nevada, Arizona, Utah and New Mexico. By 1910 U.S. mines were producing three-fifths of the world's copper output. In the early twentieth century large-scale mining was also introduced in Mexico, Chile and Canada, and later in Zaire, Zambia, Union of South Africa, Australia, Philippines, Papua New Guinea (PNG), U.S.S.R., and several Western European countries.

The shift from relatively small-scale underground mining to large-scale, open-pit mining with ore extraction of 50,000 to 100,000 tons per day determined the size of copper mines in the U.S. and in most other countries of the world. Large amounts of capital and a variety of technical skills are required to mine and process low-grade ore. Although there were once 3,000 copper mines in the U.S. alone, by 1923 eight large firms produced 64 percent of world output. Large-scale underground mining of strataform deposits in the African Copper Belt also replaced the small initial mines in this area.

WORLD RESERVES AND RESOURCES

Minerals in the earth's crust that are potentially extractable are classified in various ways, but this book is mainly concerned with three concepts as defined by the U.S. Bureau of Mines (BOM), namely, reserves, the reserve base, and resources of copper. Reserves are those materials identified and estimated by specified minimum physical and chemical criteria related to current mining practices that can be economically extracted at the time of determination. The reserve base includes those materials that are (1) currently economic to mine (reserves); (2) marginally economic (marginal reserves), and (3) currently subeconomic but with reasonable potential for becoming economically available within the next few decades. All the materials in the reserve base have been identified in that their location, grade, quality and quantity are known or estimated from specific geological evidence. Mineral resources is a broader concept that includes, in addition to identified resources, inferred resources based largely on the geologic character of deposits but for which there may be no samples or measurements. Resources include only those materials for which economic extraction is currently or potentially feasible.[6]

World copper reserves and the reserve base (as of 1985) are estimated to be 340 million mt and 509 million mt, respectively (Table 1.1). About 32 percent of the world reserve base is located in developed countries; 59 percent in developing countries; and 9 percent in Eastern Europe. World resources of copper are estimated at 2.3 billion mt, of which 1.6 billion mt are land-based and 0.7 billion mt are in deep-sea nodules. The estimate of copper in deep-sea nodules is much less reliable than that for land-based resources for two reasons. First, lifting and extracting seabed nodules is not

Table 1.1
World Copper Reserves and Reserve Base
(million metric tons)

	Reserves	Reserve base (includes reserves)
North America		
U.S.	57	90
Canada	17	32
Mexico	17	23
Other	1	15
Total	90	159
South America		
Chile	79	97
Peru	12	32
Other	3	12
Total	95	140
Africa		
Zaire	26	30
Zambia	30	34
Other	4	7
Total	60	70
Asia		
Philippines	12	18
Other	14	19
Total	25	35
Oceania		
Australia	8	16
PNG	6	14
Other	1	4
Total	15	35
Europe		
Eastern	n.a.	45
Western	n.a.	25
Total	50	70
World Total	340	509

NOTE: Totals may not add because of rounding.

Source: Bureau of Mines (1985b) "Copper," Mineral Facts and Problems, U.S. Department of Interior, Washington, p. 7; and Bureau of Mines (1985a), Mineral Commodity Summaries, U.S. Department of Interior, Washington, p. 41.

yet economical, although it is likely to become so in the next century. Second, sufficient exploration may reveal much larger quantities of nodules than have been discovered thus far.

World copper consumption has been growing at just under 2 percent per year. Assuming this rate of growth continues, the reserve base of 509 million mt would be sufficient to satisfy demand for another thirty-six years. World resources estimated at 2.3 billion mt would be sufficient to supply world demand for nearly 90 years, assuming a 2 percent rate of growth. Although it is not possible to estimate the rate of growth in demand for the next century, it seems unlikely the world will exhaust its copper resources for many decades to come.

Actually, the consumption life of an exhaustible mineral has little practical significance. Resources are never fully exhausted since increasing scarcity results in higher prices which in turn moderates demand. Higher prices induce the use of substitutes and stimulate R&D for both substitution and conservation in use. At the same time, higher prices stimulate the search for new reserves and make profitable the production of lower-grade deposits. Thus as reserves are depleted, dynamic forces are set in motion that prevent exhaustion of a mineral. As is discussed in Chapter 3, there are several substitutes for copper in nearly every use, and advances in materials technology are continually producing new substitutes and conservation methods.

WORLD PRODUCTION AND CONSUMPTION

In 1983 annual world mine copper capacity totaled 10.3 million mt, but world production was only 8.1 million mt. The difference between capacity and production is largely due to excess capacity in developed countries; there was much less excess in developing countries. In this same year 28 percent of world mine output was in the developed countries; 49 percent in the developing countries; and 22 percent in the centrally planned economies (the Soviet Bloc plus China). World smelting capacity and production are somewhat higher than mine capacity and production because some of the copper smelted is produced from scrap (secondary copper). Refining capacity and production are still higher because some secondary copper is refined with primary (from ores) (see Table 1.2). Developed country smelting and refining capacity is

Table 1.2
World Copper Capacity and Production, 1983
('000 metric tons)

	Mine		Smelter		Refinery	
	capacity	production	capacity	production	capacity	production
North America						
Canada	675	625	700	385	675	544
U.S.	1,780	1,038	1,884	967	2,822	1,584
Other	355	251	170	75	145	75
Total	2,810	1,914	2,754	1,447	3,642	2,203
South America						
Chile[a]	1,290	1,257	1,120	1,058	900	833
Peru[a]	43	322	400	258	250	191
Other	40	24	70	10	100	57
Total	1,763	1,603	1,590	1,326	1,260	1,081
Europe						
Market economies[b]	1,420	324	1,000	720	1,560	1,505
Centrally planned economies	1,880	1,511	2,200	1,746	2,116	1,826
Total	2,300	1,835	3,200	2,466	3,676	3,331
Africa						
Zaire[a]	600	535	500	467	250	227
Zambia[a]	800	563	630	581	800	575
Other	340	321	330	299	190	173
Total	1,740	1,419	1,460	1,347	1,240	975

Table 1.2 (cont.)

	Mine		Smelter		Refinery	
	capacity	production	capacity	production	capacity	production
Middle East and Asia						
Japan	50	46	1,440	1,062	1,220	1,092
Centrally planned economies	430	319	320	228	320	302
Other	740	525	560	286	380	282
Total	1,220	890	2,320	1,576	1,920	1,676
Oceania						
Australia	260	256	190	180	210	199
PNG	200	183	--	--	--	--
Total	460	439	190	180	210	199
Total Market Economies	7,880	6,270	8,994	6,368	9,512	7,316
Total Centrally Planned	2,310	1,830	2,520	1,974	2,436	2,128
World Total	10,290	8,100	11,514	8,342	11,948	9,444

a Includes electrowinning capacity at smelter or refining level, leaching at mine level, and secondary refining and smelting production and capacity.

b Includes Yugoslavia.

Note: Totals may not add because of rounding.

Source: Bureau of Mines (1985b) "Copper," Mineral Facts and Problems, Bulletin 675, U.S. Department of Interior, Washington, p. 5.

considerably larger than mine capacity since a portion of the mine output of developing countries is processed in the U.S., Western Europe and Japan.

World consumption of refined copper totaled 9.1 million mt in 1983; this amount included about 1.2 million mt of secondary copper. The developed countries accounted for about 64 percent of total world consumption; the Soviet Bloc and China about 26 percent; and the developing countries about 10 percent. However, it is expected that consumption will grow more rapidly in the developing countries than in the rest of the world. The developing countries have been growing somewhat faster in terms of per capita income than the rest of the world, and a larger portion of their income is expected to be be spent for durable goods containing copper.

STAGES IN PRODUCTION

As is the case with most minerals, the copper industry is divided into major production stages: (1) exploring; (2) mining and milling; (3) concentrating; (4) smelting; (5) refining; and (6) semifabricating. Some copper companies operate at only one or two stages, while most large firms are integrated through all six stages.[7]

Exploration

In contrast to methods employed in the last century by prospectors who depended on surface observations and samples gathered with pick and shovel, exploration has become highly complex using sophisticated technology. Subsurface deposits are located by "telegeological" or "remote sensing" techniques using aircraft or satellites; by chemical analysis of elements found in rocks, soil or streambed sediments; and by electronic equipment that can detect mineral characteristics by measuring magnetic and conductivity conditions and radioactivity. For example, the large Ok Tedi copper/gold deposit in PNG was discovered through analysis of streambed samples taken many miles from the mountain where the deposit is located. Evidence of mineral deposits has been found by LANDSAT imagery provided by satellites and by side-looking radar or photography where conventional methods are inadequate. Modern exploration is carried on by teams of experts that include geologists, geochemists, and geophysicists.

There are several stages in exploration, each having a specific goal. The goal may be to find a mineable deposit of a specific mineral or to find deposits of unspecified minerals; or it may be to learn more about a known deposit. The search for new deposits (called "grass roots" exploration), usually begins with a review of the geological and mining literature on an area, followed by a variety of aerial and ground reconnaissance and mapping designed to identify one or more potential prospects. Once a prospect has been identified, the area is drilled to obtain samples for assaying and metallurgical testing. If the sample shows ore suitable for mining, intensive exploration by systematic close-space drilling and collecting of bulk samples is undertaken to estimate the volume and location pattern of ore in the deposit. Until reserves are measured and their location known, it is not possible to determine whether mining would be profitable.

Feasibility Study

Before a decision is made to construct mine and processing facilities, a feasibility study must be made based on information on the orebody, estimates of capital and operating costs, and projections of product prices. For a large mine, a feasibility study may cost $25 million or more, and require preparation of engineering data covering all aspects of the project--the mine and plant, infrastructure, and facilities that limit environmental damage. All the elements of the cash flow analysis must be estimated and the cash flow simulated from projected information to determine the internal rate of return (IRR) from the project under various product price, cost and tax assumptions. A financial plan, including an estimate for debt financing, is contained in the feasibility study. On the basis of the feasibility study the investor decides whether the probability-adjusted rate of return on equity capital is sufficient to warrant an investment.

Mine Construction

Construction of large copper mines in both

developed and developing countries is usually undertaken by international engineering and construction firms, such as Bechtel Civil and Minerals, Inc., and Morrison and Knudsen. Modern open-pit copper mines require investment of several hundred million dollars and three to five years to construct. The several components of construction--mine preparation, crushing mills, concentrators, highways and/or railroads, utilities, and facilities for mine waste disposal--must be coordinated so that operations can begin on schedule. Even a loss of a few months can significantly increase construction costs due to interest on borrowed funds.

Mining and Milling

Most porphyry copper mines are open-pit operations. The deposit is drilled and blasted and the ore hauled to crushing mills in large trucks (with capacities running from 60 to 200 tons) or ore trains. Open-pit copper mining tends to be the rule in the Western Hemisphere and in the Far East, and there is a ring of porphyry copper deposits circling the Pacific Ocean. Some strata-bound deposits are also mined by open pit methods as is the case for large mines in Zaire, while underground mining of strata-bound deposits is predominant in Zambia. Underground mining is more costly and, therefore, requires a higher copper content to be profitable. In underground mining the ore is drilled, blasted and moved through passageways to a centrally located shaft through which it is lifted to the surface. In the large Zambian underground mines the ore goes through a primary crusher several thousand feet below the surface and is then moved by conveyor belt to a secondary crusher on the surface.

Concentrating and Leaching

After the ore is mined, the copper-bearing minerals must be separated from the rock and the copper extracted from the minerals. The methodology used depends upon the type of minerals mined. In the case of most sulfide minerals, the ore is crushed and finely ground so that individual mineral grains are separated from the rock and the copper-bearing minerals removed by flotation to produce concentrate. Copper concentrate usually contains from 22 to 32 percent copper metal by weight. In the case of other types of minerals (including copper oxide, carbonates and silicates), other methods must be used to separate the copper-bearing minerals.

Leaching is used in Zaire and in certain deposits in Arizona and Chile. There are several methods of leaching, but the basic procedure is to treat the ore with a chemical solution until most of the copper has been collected in the solution. Copper is then extracted from the solution by precipitating it on scrap iron, or by recovering it directly as copper metal in electrolytic cells. In the latter method, electrowon cathode is produced by electroplating copper cathodes directly from the leach liquor. In the case of in situ leaching, a sulfuric acid solution is allowed to flow through the deposit and collects as leached liquor beneath the deposit, thereby eliminating mining. This method is only practical for certain deposits. Hydro-metallurgical methods involving chemical extraction by leaching and electrowinning of the leached solution avoid the environmental problems associated with pyrometallurgical methods in which concentrates are roasted and smelted.[8]

Smelting

Most copper is produced by smelting concentrates in a furnace to produce blister that is 97.0 to 98.5 percent pure. (Refined copper is 99.8+ percent pure.) Today the most commonly used furnaces are (1) the reverberatory; (2) the electric; and (3) the flash. The reverberatory furnace was widely used in the U.S., South America and Africa early in this century, but is being phased out in the U.S. because it is energy inefficient and does not meet environmental standards. The reverberatory furnace produces a large volume of sulfur dioxide (SO_2) gas in concentrations too low for efficient sulfur recovery. The flash smelting process was developed in Finland in the early 1950s. Copper sulfide concentrates are smelted by burning a portion of the contained iron and sulfur with preheated air or oxygen-enriched air injected into the furnace. In this process SO_2 is sufficiently concentrated to produce sulfur. A copper matte is produced which is transferred to a converter where air flowing through the matte burns off the sulfur, oxidizes the iron for removal in slag, and yields blister copper. The blister copper is then melted in a furnace and cast into copper anodes for electrolytic refining. The anodes and cathodes (thin copper starting sheets) are suspended in tanks containing a solution of copper sulphate and sulfuric acid. An electric current is passed through the solution which dissolves copper from the anodes and deposits it in refined form on the cathodes.

Copper concentrates are also smelted in electric furnaces to produce blister copper which is then electrolytically refined. Currently, there are only two electric furnaces in operation in the U.S. for smelting concentrates. They are more widely used in the production of secondary refined copper from scrap.

Most mining complexes do not include copper refineries. There is a tendency to locate refineries in highly industrialized areas where the products are used, or near ports where they are shipped abroad. Refined copper is produced in five principal shapes: (1) cathodes, (2) wirebars, (3) continuous cast rods (CCR), (4) billets, and (5) cakes. Wirebars, which may weigh from 200 to 300 pounds, have in the past been used mainly for the manufacture of copper wire and rod. However, wirebars are rapidly being displaced by high purity cathode copper, which has become the most widely traded primary copper form, and by CCR. CCR measures 5/16 of an inch in diameter and is produced in coils that range from 5,000 to 15,000 pounds. Copper billets are solid cylindrical shapes and are used mainly for seamless copper tubing. Copper cakes are large slabs weighing more than thirty tons and are used to make plates, sheets, strips and bars.

Secondary Copper

A substantial portion of copper products are recycled from scrap. There are several qualities of scrap, the highest being new scrap from the workings of refined copper. There are different grades of old scrap contained in discarded materials and equipment. New scrap can be refined directly, while old scrap usually requires smelting to remove impurities. New scrap and some high-grade old scrap are used with mine copper in primary refineries, but secondary refineries use both new and old scrap to produce refined copper that is competitive with primary copper for most uses. In 1983 about 25 percent of the refined copper consumed in the U.S.

was produced from scrap. Scrap is also used in brass mills and for other direct consumption without being refined. There is, therefore, an important market for scrap competing with mine copper (Bureau of Mines, 1985b, p. 40).

Semifabrication

Most copper companies in the U.S., Western Europe and Japan are integrated into semifabrication. The most important semifabricating plants are wire mills and brass mills. Wire mills use primary refined copper exclusively, while brass mills use a large amount of copper scrap. Other semifabricated products include copper powder and castings produced by foundries.

ORGANIZATION OF THE INDUSTRY

In 1980 the bulk of the mine output of the market economy countries was produced by about twenty multinational corporations (MNCs) and a half-dozen state-owned mining enterprises (SMEs) in developing countries. Most of the firms are integrated through refining and some of the MNCs through semifabricating. Less than 12 percent of mine copper is produced by private mining firms with an annual output of under 100,000 mt, and most small mining companies produce only ores or concentrates. Over 40 percent of mine output is controlled by MNCs and about 35 percent by SMEs. The share of smelting and refining capacity is somewhat larger for the MNCs since a portion of the output of developing countries is processed abroad.[9]

U.S., British and Canadian multinationals dominate the private production of copper. The largest U.S. multinationals are ASARCO, Kennecott (a subsidiary of Standard Oil of Ohio), Newmont, and Phelps Dodge. The largest British multinational is Rio Tinto Zinc (RTZ), while the largest Canadian Multinationals are INCO and Noranda. In 1982 the companies just named accounted for about 30 percent of copper output of market economy countries in 1982. The largest SME producers are CODELCO (Chile), ZCCM (Zambia), and Gecamines (Zaire), which in 1982 accounted for over one-third of the output of the market economies.

There have been substantial changes in the ownership of world copper capacity during the past two decades. The major change has been the growth of SMEs in developing countries, mainly due to the nationalization of properties owned by MNCs in Chile, Peru, Zaire and Zambia. At the same time, there has been an expansion of copper production by multinational and private domestic firms in Australia, Canada, Indonesia, PNG and Peru. Low copper prices for the first half of the 1980s have resulted in a sharp decline in U.S. production, while Chilean production (mainly CODELCO) has risen rapidly.

INDUSTRIAL USES

The largest use of copper is in electrical supplies and equipment, accounting for nearly 60 percent of U.S. copper demand in 1983. Copper wire is used in the manufacture of electrical motors and generators, power transmission lines, housing and industrial wiring, and other types of electrical wiring. Telecommunications accounts for 12 to 15 percent of annual copper consumption in the U.S., but in recent years substitution of aluminum, optic fibers, and other materials has substantially reduced this demand.

Construction is the second largest category in which copper is used, accounting for about 19 percent of total U.S. demand. Construction uses include roofing, gutters and downspouts, and heating and cooling tubing. A third use of copper is in nonelectrical machinery, air conditioning equipment, and farm machinery--accounting for about 9 percent of U.S. demand. An important but declining demand is for use in transportation equipment--autos, railroads, marine vessels and aircraft--constituting about 7 percent of U.S. demand in 1983. A final important demand category is for ordnance, including ammunition and a variety of weapons. The relative importance of copper in military goods has declined in recent decades so that by 1983 it accounted for only 1 percent of U.S. demand. There are a large number of other uses, such as chemicals, kitchen utensils, and coins, but in 1983 the sum of these uses was less than one-half of 1 percent.[10]

THE WORLD MARKET

The world market for copper is not homogeneous, and there are about forty types of copper materials traded. Several grades of refined copper are traded on the commodity exchanges, the most important of which are the London Metal Exchange (LME) and the New York Commodity Exchange (COMEX). Electrolytic wirebar and cathode are traded on the LME, while these two plus high conductivity fire-refined copper are traded on the COMEX. There are also commodity exchanges in other countries that deal in refined copper. Most of the trade on commodity exchanges is in contracts for future delivery, so they serve mainly as markets for hedging. Most refined copper is sold under contracts between producers and consumers either at a price established by the producer--the producer price--or at a reference price that varies with prices quoted on the LME or COMEX. Producers in the U.S., Canada and certain other developed countries use the producer price system, while copper produced in developing countries is generally sold at a price tied to the LME.

Various types of copper are also traded in merchant or dealer markets. Independent dealers in this market arrange purchases of refined copper for consumers who do not want to buy under contract at the producers price; or for producers that want to sell some output on the open market. Secondary copper and concentrates are also sold on the merchant markets.

Since a number of mines do not have smelters, there is a substantial domestic and foreign market for concentrates. Normally, producers of concentrates negotiate contracts with smelters and refineries to process the concentrates into refined copper. Contracts may provide for either the outright sale of concentrates, or custom smelter/refiners may charge a processing fee, with the refined metal returned to the seller of the concentrates for marketing. There is also a market for copper scrap which may be sold under contracts between scrap dealers and smelter/refiners, or sold in the merchant markets. Although the prices of all types of copper sold in different markets throughout the world tend to move together, there are frequently significant differences in prices for the same grade of copper. This is due mainly to differences in the terms of contracts under which it is sold. Recent developments in the world market for copper are discussed in Chapter 4.

NOTES

1. Native copper was used by the American Indians and by East Africans long before any contact with Europeans.

2. In China iron was first alloyed with carbon to make steel and later introduced in Europe by the Arabs.

3. For surveys of early copper mining, see Joralemon (1973), Chapters 1 and 2; and Prain (1975), Chapters 1 and 2.

4. The source of American Indian copper was a mystery until the early 1840s. For a fascinating account of discovery and development of Lake Superior copper, see Joralemon (1973), Chapter 3.

5. It has been estimated that between 1851 and 1880 Chile produced 1.2 million mt of copper, or 40 percent of world output. See Corporacion del Cobre (1975), p. 25. Today Chile produces about 1 million mt of copper per year, over half of it from two large mines.

6. For definitions of the reserve base and resources, see Bureau of Mines (1985a), pp. 180-83.

7. For a more comprehensive description, see John W. Whitney, "Physical Characteristics of the Copper Industry," in Mikesell (1979), Chapter 2.

8. For a discussion of leaching methods, see Mikesell, (1979), pp. 60-63; and Bureau of Mines (1985b), p. 9.

9. For a description of uses of copper see Bureau of Mines, (1985b), pp. 10-13.

10. These figures are only approximate and have been derived from various sources, including Radetzki, (1985), Chapter 2; United Nations, (1980), Chapter 2; and Carman, (1985), pp. 119-124.

REFERENCES

Bureau of Mines (1985a) Mineral Commodity Summaries, U.S. Department of Interior, Washington.

________ (1985b) "Copper," Mineral Facts and Problems, Bulletin 675, U.S. Department of Interior, Washington.

Carman, John S. (1985) "The Contribution of Small-Scale Mining to World Mineral Production," Natural Resources Forum, 9, 119-24.

Corporacion del Cobre (1975) El Cobre Chileno, Editorial Universitario, Santiago, Chile.

Joralemon, Ira B. (1973) Copper, Howell-North Books, Berkeley.

Mikesell, Raymond F. (1979) The World Copper Industry, Johns Hopkins University Press for Resources for the Future, Baltimore.
Prain, Sir Ronald (1975) Copper: The Anatomy of an Industry, Mining Journal Books, London.
Radetzki, Marian (1985) State Mineral Enterprises, Resources for the Future, Washington.
United Nations (1980) Mineral Processing in Developing Countries, New York.

Chapter 2

RECENT CHANGES IN COPPER PRODUCTION, TRADE AND INDUSTRIAL ORGANIZATION

In 1925 the U.S. produced over half the world's output of mine copper, and the U.S. and Western Europe together produced 60 percent of world output. Refined production was even more concentrated in the U.S. and Western Europe since most of the world's copper was refined in these areas. After World War II U.S. and Western European shares in the copper output of the market economies declined rapidly in favor of developing countries in Latin America and Africa, and of Canada and Australia. By 1960 the mine output of the developing countries was nearly equal to that of the developed countries, and by the 1980s the output of developing countries was substantially higher than that of developed countries (see Table 2.1). By 1984 U.S. mine output was only 13 percent of world output and the U.S. and Western Europe together produced less than 17 percent of world output. The output of the U.S.S.R. was negligible in 1925, but by 1980 nearly equaled that of the U.S.; in 1984 the output of the communist countries (including China) was 24 percent of world output. The share of refined copper output of the major industrial countries has also been declining. In 1981 refined copper output of the U.S., Western Europe and Japan (which produces little mine copper) was 61 percent of the total market economy output. By 1984 this share had declined to 54 percent, largely due to cutbacks in U.S. mine and smelter production.

Between 1960 and 1980 mine copper output of the developing countries grew by 82 percent, while that of the developed countries grew by only 54 percent. Most of the increase in copper producing capacity in developing countries was created by MNCs. The major investments were made in Chile, Peru, Indonesia, PNG, Zaire and Zambia, which countries accounted for

Table 2.1
Mine Copper Production by Major Country and Region,
Selected Years
('000 metric tons)

	1925	1938	1960	1976	1980	1984
U.S.	761	506	994	1,461	1,184	1,090
Canada	51	267	399	733	719	714
Latin America	359	448	798	1,332	1,623	1,902
Chile	243	351	533	1,008	1,071	1,294
Peru	37	38	182	221	368	370
Other	79	59	83	103	184	236
Western Europe[a]	120	160	111	299	285	323
Asia[b]	76	151	209	482	496	657
Africa	109	395	978	1,474	1,363	1,390
South Africa	8	11	47	197	213	216
Zaire	90	124	303	446	461	490
Zambia	2	255	578	711	598	567
Other	9	5	50	120	91	117
Oceania	13	20	117	397	391	402
Australia	12	20	117	219	244	238
PNG	1	--	--	177	147	164
Total market economies	1,489	1,947	3,596	6,178	6,061	6,477
Communist countries	7	115	548	1,700	1,825	2,057
U.S.S.R.	5	115	461	1,134	1,134	1,201
China	1	--	40	150	166	200
Poland	--	--	11	268	344	405
Other	1	--	36	148	181	251

Table 2.1 (cont.)

	1925	1938	1960	1976	1980	1984
Developing countries (noncommunist)	470	881	1,849	3,186	3,363	3,853
Developed countries[c]	1,019	1,066	1,747	2,992	2,698	2,625
Total World	1,496	2,062	4,144	7,878	7,885	8,534

a Includes Yugoslavia.

b Includes Turkey.

c Includes South Africa, Spain and Yugoslavia.

NOTE: Totals may not add because of rounding.

Source: Bureau of Mines, Minerals Yearbook (various issues), U.S. Department of Interior, Washington; and American Bureau of Metal Statistics, Yearbook (various issues), New York; and Metallgesellschaft, Metal Statistics (various issues), Frankfurt.

the bulk of Third World increases in capacity. During a relatively short period from 1968 to 1974, about two-thirds of the copper producing capacity of developing countries was nationalized. The nationalized mines included most of the Chilean capacity; one-fourth of the Peruvian capacity; and all the capacity of Zaire and Zambia. Copper output in Peru, Indonesia, and PNG is dominated by the MNCs, but foreign investment in copper capacity in developing countries has been low since the mid-1970s.

SHIFTS IN THE LOCATION OF WORLD PRODUCTION

We may consider shifts in world copper production from the standpoint of (1) geographical locations; (2) output shares between SMES and MNCs; or (3) between groups of countries classified as developed, developing and communist. With regard to (4), I have noted the substantial shift in the share of production in favor of developing countries during the post-World War II period, and the growth in the share of communist countries from almost negligible in the 1930s to nearly one-fourth of world output in the 1980s. In this section I shall review the important geographic shifts in the shares of copper production among countries.

Changes in U.S. Mine Output

One of the most important changes in world copper production has been the decline in the share of U.S. output. Prior to the 1980s, this decline occurred gradually throughout the post-World War II period until 1981. The U.S. share of mine production in the market economies was 36 percent in 1950, about 28 percent in 1960, and about 24 percent over the period 1976-1981. Thereafter, mine copper output fell sharply, both absolutely and relative to the total output of market economies. Between 1981 and 1984, U.S. mine output declined by 345,000 mt and its share fell from nearly 24 percent to less than 17 percent.

The decline in the U.S. share of world production prior to the 1980s is readily explained by the development of mines in new producing areas with an abundance of higher grade deposits. After the mid-1970s, both U.S. and total output of the market economies stagnated due to the slow growth in demand and decline in real copper prices. The sharp fall in U.S. output after 1981 calls for special explanation.

The decline in U.S. copper production after 1981 took the form of cutbacks in mine output and the closure of mines. The fundamental reason for both was falling real prices in the face of rising real costs. Most U.S. mining firms were experiencing losses on their copper operations. Losses could be trimmed by shutting down high-cost mines or by reducing production of lower grade ores within mines. During the period 1982-1985, U.S. copper mines with listed capacities totalling over 600,000 mt were shut down, and much of this capacity is likely to remain closed (Seidenburg, 1984 and 1985). For example, Anaconda (owned by Atlantic Richfield), which had been the third largest U.S. copper producer, was nearly out of the domestic copper business by 1985.

Copper production in most other producing countries has been either rising or has not significantly changed since 1981, and world output has been rising moderately. Why did the world's largest producer before 1982 experience such a sharp decline in output? Two factors are largely responsible. First, the 36 percent rise in the real exchange value of the dollar[1] between 1980 and 1985 raised U.S. production costs relative to those of other countries. (U.S. costs were already higher than average world costs prior to the 1980s.) Second, a high proportion of U.S. smelters were old and inefficient and did not meet Environmental Protection Agency (EPA) standards. U.S. copper firms were under strong pressure by EPA to replace old reverberatory smelters in order to meet these standards. Given the low profitability of the copper mining industry, it did not pay to modernize or replace smelters and a number were shut down. This meant that either the mine was closed or the concentrates were shipped to other smelters.

Additional shutdowns of inefficient and environmentally substandard smelters are likely to occur in the second half of the 1980s. One forecast is that by the end of the decade about 20 percent of U.S. concentrate output will be shipped to Japan for smelting and refining; about 12 percent will be produced by leach-electrowinning; and the remaining 68 percent will be treated domestically by smelting and refining (Seidenburg, 1985, p. 22). This suggests that half the 1984 smelter capacity will have been abandoned by the end of the decade.

However, new copper processing facilities are being established in the U.S., and some old ones modernized. A significant rise in the price of copper would undoubtedly spur new investment in processing facilities that would meet environmental standards.

Canada

Canadian mine production rose very rapidly during the post-World War II period, reaching a peak of 820,000 mt in 1974. Thereafter, output declined by about 10 percent with a further decline during the 1982-83 recession, but recovered to over 700,000 mt in 1984. Average Canadian production costs are somewhat lower than the U.S. average and many Canadian deposits contain significant amounts of gold as a byproduct. Nearly half of Canada's copper concentrates are processed abroad and environmental controls are not as stringent as those in the U.S. (This is largely true because mines tend to be located considerable distances from populated areas.)

Chile

Chile became the largest mine copper producer in the world in 1982 and continued to increase its output through 1984 when it reached nearly 1.3 million mt. This expansion has been led by CODELCO, which has maintained a large investment program while investment in the world copper industry was generally declining. Privately-owned mines, including EXXON'S Disputada mine, also increased output. Chile's production costs are among the lowest in the world. Plans are being made for investments by copper MNCs for development of new deposits.

Peru

Nearly 70 percent of Peru's output is produced by Southern Peru Copper Corporation (SPCC) (owned by four American mining companies), and the remainder is produced by two SMEs one of which, Centromin, was owned by U.S. investors before it was nationalized in 1974. Peru's mine output rose rapidly during the late 1970s to a peak of about 400,000 mt in 1979, but declined by about 10 percent during the first half of the 1980s. In the 1970s, the growth in output was mainly due to the completion of SPCC's large Cuajone mine. However, SPCC's mines are being depleted and without additional investment Peru's output is likely to continue to decline. SPCC hopes to reach an agreement with the government to develop the Quellaveco deposit which is adjacent to its other mines. (SPCC owned Quellaveco before it was expropriated in 1971). The foreign investment climate in Peru is currently so poor that large new private investment seems unlikely. The government has been seeking foreign capital for development of other previously expropriated orebodies, but such financing is unlikely to become available due to Peru's external debt problems.

Zaire and Zambia

Zaire's annual copper output reached 500,000 mt in 1974 and, except for short-run decreases arising mainly from adverse political conditions, it has maintained that level through 1984. On the other hand, Zambia's production, which reached a peak of over 700,000 mt in 1976, has declined during the 1980s and averaged only 557,000 mt in 1983-84. The copper industries of both countries have been handicapped by transportation problems in delivering their products to ocean ports; both countries have also suffered from shortages of equipment and supplies, partly as a consequence of foreign exchange difficulties. Zambia's foreign exchange situation is more serious than that of Zaire. Despite considerable investment in Zambia's copper industry, financed in part by foreign loans, its output has been well below capacity and labor productivity has declined.

Other Countries

Australia, Indonesia, Iran, Mexico, Spain, Sweden and Yugoslavia have increased their output significantly since the mid-1970s. Multinational mineral companies played an important role in the expansion of output in Australia, Indonesia and Spain. Philippine output rose rapidly until 1979 and then declined, especially after 1981. Among the communist countries, the largest percentage gains in copper output occurred in China, Mongolia and Poland.

CHANGES IN OWNERSHIP AND CONTROL OF THE WORLD INDUSTRY

The most important changes in ownership and control of the copper industries in market economy countries over the past fifteen to twenty years have been (1) the growth in SMEs in developing countries and (2) the takeover of a number of large multinational mining firms by petroleum companies and other nonmining firms. A third and somewhat less important development has been the increase in domestic ownership and control of subsidiaries of MNCs in Australia and Canada. A fourth development has been the minority participation of governments in the ownership of subsidiaries of MNCs in developing countries. Table 2.2 shows the major Third World SMEs and their 1984 mine output. Nearly 85 percent of this output was produced by SMEs formerly owned by MNCs.

Table 2.3 shows the major Third World copper mines owned and controlled by MNCs and their output in 1984. The output of this partial list of copper mines was 610,000 mt, as contrasted with 2,569,000 for the SMEs listed in Table 2.2. However, Tables 2.2 and 2.3 exclude a number of mines in which SMEs have a substantial interest with the remainder held by private investors. For example, the Mexican government has 44 percent of the equity in Mexicana del Cobre, with a capacity of 180,000 mt, but the mine is under private control. Malaysia's Mamut mine is owned 49 percent by the government, but control is in private hands. Some 40 percent of the equity in Zambia's ZCCM is owned by foreign investors, but the government has full control.

Most of the copper output controlled by MNCs is in developed countries and in many cases it is difficult to determine the locus of control because mining companies are often owned by more than one MNC and, in addition, there is frequently substantial ownership by small private shareholders. From the standpoint of world ownership and control of production, the most important MNCs are ASARCO, RTZ, Phelps Dodge, Newmont, and Anglo-American of South Africa (AAC). RTZ has important copper investments in Australia, the U.S., PNG, South Africa, Canada and Spain. ASARCO has large copper investments in Canada, Peru and Australia as well as in the U.S. Newmont has large investments in the U.S., Canada, Namibia and South Africa. AAC has large investments in South Africa, Canada, Australia and the U.S. As of 1980, large U.S. copper companies controlled about 900,000 mt of annual production outside the U.S., and major foreign MNCs controlled about 400,000 mt of annual production outside the U.S.

Table 2.2
Mine Copper Production of
Major Third World State Mining Enterprises, 1984
('000 mt)

Brazil	
Caraiba Metais	40
Chile	
CODELCO*	1,053
Empresa Nacional de Mineria	169
India	
Hindustan Copper	41
Mexico	
Mina de Cananea	49
Peru	
Centromin*	50
Mineroperu	27
Zaire	
Gecamines*	490
Zambia	
ZCCM*	523
Yugoslavia	
R.T.B. Bor	127
Total	2,569

* Formerly owned by MNCs.

Sources: American Bureau of Metal Statistics (1984) Yearbook, New York; and CIPEC (1986) Quarterly Review, Paris, January-March.

Takeovers of Major Mining Companies

During the late 1970s and early 1980s a number of mining companies were acquired by large petroleum or other companies not previously in the mining business (a partial list is shown in Table 2.4). From the standpoint of increasing or maintaining

Table 2.3
Copper Mines of MNCs in Third World Countries, 1984
('000 mt)

Mine and country	MNC with controlling equity interest	Output
Bougainville Copper Ltd.--PNG	Rio Tinto Zinc	165
Freeport Indonesia--Indonesia	Freeport-McMoRan	84
BCL--Botswana	AAC/AMAX	22
Minera Disputada de las Condes--Chile	EXXON	62
SPCC--Peru	ASARCO	252
Tsumeb Corp.--Namibia	Newmont	25
Total		610

Source: American Bureau of Metal Statistics (1984), Yearbook, New York.

copper producing capacity, the takeovers have had two opposing effects. First, the large resources of the acquiring companies have provided financing for increased investment. This was undoubtedly a factor in the expansion of the Disputada mine in Chile following its purchase by EXXON. On the other hand, the control of formerly independent mining firms by nonmining firms has in some cases resulted in closing or selling unprofitable mines and reducing exploration activities. The best example of this is the withdrawal of Anaconda from the mining business after it was taken over by Atlantic Richfield in 1977. Kennecott (which was the second largest copper mining firm in the U.S.) cut back its operations to a greater degree than did Phelps Dodge and ASARCO (which have remained independent), and in 1986 sold two of its copper mines, one to Phelps Dodge and the other to ASARCO. Independent mining firms tend to be controlled by managers devoted to the mining industry, while those controlled by conglomerates are more likely to divest mining

Table 2.4
Copper Mining Companies Acquired by Nonmining Firms

Acquiring firm	Date acquired	Acquired mining company	Copper production 1981 ('000 mt
Atlantic Richfield	1977	Anaconda (U.S.)	136[a]
Louisiana Land & Exploration	1977	Copper Range (U.S.)	39
EXXON	1978	Disputada (Chile)	39
Standard Oil of Indiana	1979	Cyprus (U.S.)	107
Standard Oil of Ohio	1981	Kennecott (U.S.)	338
Standard Oil (Mobil)	1980	Falconbridge (Canada)	47
McMoRan[b]	1981	Freeport Indonesia (Indonesia)	63
Total			769

a Includes half of capacity of ANAMAX mines owned jointly with AMAX.
b Freeport Minerals merged with McMoRan to form Freeport McMoRan in 1981. Freeport Minerals began work on the Ertsberg (Indonesia) mine in 1967; Freeport Indonesia is the operating subsidiary.

Source: American Bureau of Metal Statistics (1984), Yearbook, New York.

assets when they are no longer profitable relative to other conglomerate activities.

OUTLOOK FOR CAPACITY

Over the long run, say, twenty-five years, copper producing capacity will tend to grow with demand so that projecting capacity is in large measure a matter of projecting consumption. But for periods of under ten years, planning and construction time puts an upper limit on new capacity expansion. Also, overcapacity and low prices will keep capacity growth less than consumption growth. Account must also be taken of depletion and the need to replace existing processing facilities since, in the absence of any new investment, capacity declines over time. Following a period of shut down capacity, some mines will be restored to production while others will remain permanently closed. In some cases it may be difficult to determine which mines are permanently closed. Much depends upon how well the mines are maintained. When a mine has been closed for several years and all equipment liquidated, and perhaps the shafts flooded or clogged with rocks, it is unlikely ever to be restored to production.

Projections of productive capacity over a five-year period have been made regularly by various organizations, the most detailed of which are those by the Paris-based Intergovernmental Council of Copper Exporting Countries (CIPEC). Such projections are made on the basis of planned projects, depletion of existing mines, and expectations regarding the reopening of currently idled capacity. These projections can, of course, be affected by developments in demand and prices since planned projects, including those already under construction, may be delayed or accelerated in response to changes in prices. Increases can also take place in the shorter run by expanding capacity of existing mines and processing facilities.

At the end of 1984 total mine capacity of the market economies was estimated at 8.1 million mt. According to a study by CIPEC, this capacity is projected to increase to 8.8 million mt by the end of 1990--an average annual rate of growth in mine copper capacity of 1.5 percent per year (CIPEC, 1985, pp. 2-3). There are, however, substantial differences in the projected changes in capacities among countries and regional groupings. U.S. and Canadian capacities are projected to decline slightly, but their combined share of total market economy capacity is projected to decline from 32.1 percent at the end of 1984 to 28.5 percent by the end of 1990. Chile's capacity is projected to increase from 1,352 thousand mt at the end of 1984 to 1,949 thousand mt at the end of 1990, or by nearly 600,000 mt. This represents nearly 78 percent of the net increase in the mine producing capacity of the market economies over this period. Capacity increases are also projected for Australia, PNG, Zambia, Iran, the Philippines, and Brazil. However, decreases in mine capacities are projected for the countries of Western Europe, Zaire, South Africa and Indonesia (see Table 2.5).

A World Bank study (Takeuchi, et.al., 1986,

p. 126) projects much smaller increases in copper mine capacities of market economies. While CIPEC has projected an increase in total capacity of 771,000 mt between the end of 1984 and the end of 1990, the World Bank study projects an increase of only 500,000 mt. About 70 percent of this difference is accounted for by projections in Chile's mine capacity by the two sources. The CIPEC study projects an increase in Zambia's capacity; while the World Bank projects a sharp decline. Another important source of difference is found in the projections of U.S. capacity. CIPEC shows a decline of only 50,000 mt for the U.S. between the end of 1984 and the end of 1990, while the World Bank study projects a 250,000 mt decline. It is worth noting the World Bank study projects almost no change in copper mining capacity in the market economies between 1990 and 1996. Varying projections reflect differences in judgments regarding the realization of planned increases in capacities in particular countries and the number of existing mines that will be permanently closed.

The Phelps Dodge Corporation's research staff has projected the growth of non-communist copper mine capacity over the period June 30, 1985 to December 31, 1989. It estimates this capacity will increase from 7,945 thousand mt to 8,423 thousand mt, or by 478,000 mt. This increase is about the same as the World Bank projection. It is interesting to note the Phelps Dodge staff projects only a modest decline in U.S. capacity between mid-1985 and the end of 1989. As is the case with the other two projections, the vast bulk of the increase in capacity in the market economies is attributed to Chile with a modest increase for Mexico and Western Europe; capacity for most other producing countries is expected to remain the same or decline (Phelps Dodge, 1985).

The U.S. BOM has projected an average annual rise in non-communist copper mine capacity outside the U.S. of 2.6 percent between 1983 and 1990. This represents a considerably higher projected rate of growth in mine capacity for market economies than that projected by the World Bank and Phelps Dodge, but a somewhat lower rate than that projected by CIPEC (Bureau of Mines, 1985b, p. 24).

There is considerable disagreement among investigators regarding U.S. mine capacity in the 1990s. For example, Siedenburg believes that most of the U.S. capacity shut down in 1984 will not be reopened and that further permanent closures will

Table 2.5
Mine Copper Capacities 1984 and Projected 1990
('000 mt)

	End 1984	End 1990
Developed Countries		
U.S.	1,643	1,593
Canada	941	921
Western Europe[a]	400	357
Australia	256	271
South Africa	228	203
Other	48	43
Total	3,516	3,388
Developing Countries		
Chile	1,352	1,949
Zaire	705	676
Zambia	640	674
Mexico	311	388
Peru	381	380
Philippines	400	464
PNG	165	220
Iran	106	151
Indonesia	91	84
Other	390	454
Total	4,541	5,440
Total Market Economies	8,057	8,828

a Includes Yugoslavia.

Source: CIPEC (1985) Survey of Mine, Unrefined and Refined Capacities, Paris, June, p. 11.

occur in the late 1980s, leaving U.S. capacity less than 900,000 metric tons per year (mtpy) by the end of the decade (Siedenburg, 1984, pp. 22-24; and 1985, pp. 18-22). Although little new U.S. capacity is expected to be built by 1990, the amount of idled capacity restored to production will depend heavily on the price of copper.

On the other side of the ledger, neither the CIPEC nor the World Bank projections include several large planned projects scheduled to come on stream by 1990. One is the La Escondida project in Chile--owned 60 percent by Utah International, 30 percent by RTZ, and 10 percent by a consortium of Japanese companies. Construction could start by 1988 with production of 280,000 mtpy copper in concentrates by the early 1990s. A second planned project in Chile is Cerro Colorado, owned by Rio Algom (a subsidiary of RTZ), which is designed to produce 60,000 mtpy copper concentrates. A third project, the Olympic Dam copper/gold/uranium project in Australia, is designed to produce 35,000 mtpy copper. Finally, there is the Neves-Corvo project in Portugal designed to produce 65,000 mtpy copper concentrates (Commodities Research Unit, 1985, p. 11-12). Since financing has not been arranged for these projects, there is perhaps less than a 50-50 chance of completion by 1990. There are a number of other mining projects with announced startup dates before the end of this decade that are unlikely to be constucted before the 1990s (Sassos, 1986, pp. 25-28).

SMELTING AND REFINING CAPACITY

CIPEC projects smelter and chemical processing capacities will increase by 468,000 mtpy, with the principal additions in Mexico, Canada, Australia, Chile and Spain. U.S. smelter capacity is projected to decline by only 57,000 mtpy, but since the CIPEC projection was made several hundred thousand mtpy of U.S. capacity has been permanently closed and by the end of 1990, U.S. smelter and chemical processing capacity may well be under 1 million mtpy. This would mean a substantial decline in the total processing capacities for the market economies, and a shortage of capacity for processing the projected volume of copper concentrates.

CIPEC projects copper refining capacity to increase by 587,000 mtpy between the end of 1984 and the end of 1990, with over 50 percent of the increase accounted for by Chile, and another 28 percent by Zaire and Zambia. U.S. and Western European refining capacity is projected to decline moderately, while Japan's capacity is projected to remain constant. Refining capacity in Iran, the Philippines, Mexico and Canada is projected to rise. Overall, these projections suggest a shift in refining capacity from the U.S. and Western Europe (whose capacity substantially exceeds domestic mine production), to those countries that have been shipping concentrates and unrefined copper metal abroad for refining. This trend toward processing copper in countries where it is mined is likely to continue, but by 1990 there will still be a large amount of copper produced in developing countries and refined in the U.S., Western Europe and Japan. In 1984 the refinery capacity utilization rate was low in the U.S. (65 percent) and in Japan (72 percent), while in Chile and Zaire it was 98 and 90 percent, respectively. For the market economies as a whole, the 1984 average capacity utilization rate was 77 percent (CIPEC, 1985, p. 39).

MINE CAPACITY UTILIZATION RATES

Copper mine capacity utilization rates in the U.S. and Canada have been exceptionally low during the first half of the 1980s. These low utilization rates--67 percent and 63 percent respectively in 1983, and 66 percent and 75 percent respectively in 1984--arose mainly from the large number of mine closures. Between 1981 and 1985 twenty-four major U.S. mines with capacities of 887,000 mtpy were closed, of which about 30 percent are permanently closed. During the same period, eighteen major Canadian copper mines with capacities totaling 214,000 mtpy were closed, of which 16 percent are regarded as permanently closed (CIPEC, 1985, pp. 15-17). Other countries with low capacity utilization rates in 1984 were the Philippines (55 percent), Spain (72 percent), Mexico (60 percent), Yugoslavia (71 percent), and Zaire (71 percent). In the case of some of these countries, the low rates are not necessarily caused by low prices, but are due to other factors, such as strikes, transportation difficulties, and inability to obtain equipment and supplies or make repairs. Countries with relatively high utilization rates in 1984 included Australia (96 percent), Chile (95 percent), Indonesia (99 percent), PNG (99 percent), and Peru (96 percent).

WORLD TRADE PATTERNS

Annual gross exports of copper totaled over $9.0 billion in 1980-1982 (average) and of this amount $5.6 billion was exported by developing countries and $3.5 billion by developed and communist countries (World Bank, 1985, p. 16).[2] The quantity of developing country exports rose from an average of 1,874 thousand mt in 1961-1965 to 3,465 thousand mt in 1984. Industrial country exports rose between these two periods from 1,106 thousand mt to 1,596 thousand mt, while industrial country imports rose from 2,658 thousand mt to 4,301 thousand mt (Table 2.6). Although developed country imports rose substantially more than developing country imports in terms of absolute quantities, the rate of rise in developing country imports was much greater, and this trend in likely to continue.

Table 2.6
International Trade in Copper
('000 mt)

	1961-65	1971	1980	1984
Developing country exports	1,874	2,387	1,365	3,465
Developing country imports	670	1,241	2,381	1,467
Industrial country exports	1,106	1,373	1,724	1,596
Industrial country imports	2,658	3,316	4,289	4,301

NOTE: Includes refined copper, blister copper, and ores and concentrates.

Source: World Bank (1985) Commodity Trade and Price Trends, Johns Hopkins University Press, Baltimore, Tables 10, 14, and 16.

Most major developing country copper producers are copper exporters. However, of the major developed country producers of mine copper only Canada, Australia and South Africa are significant net exporters. Since some of the industrial countries, including Belgium, West Germany and Japan, have smelting and refining capacities, these countries are large exporters of refined copper although most of their concentrates are imported, mainly from developing countries. Prior to World War II, the US was the leading exporter and although although in most years during the 1960s and 1970s it was a net importer, as late as 1981 its net import reliance was only 5 percent of consumption. However, in 1983 and 1984 U.S. net import reliance as a percent of consumption was 19 and 21 percent respectively. This was due mainly to the domestic industry's inability to compete with imports, rather than a shortage of productive capacity in relation to consumption. It appears likely the U.S. will continue to be a substantial net importer of copper during the remainder of the present century.

There has been an important change in the shares of different types of copper entering into world trade during the past twenty-five years. In 1960 27 percent was blister and 63 percent refined, with ores and concentrates representing less than 10 percent of total exports. By 1984, 26 percent of world trade was in ores and concentrates (mainly concentrates), 16 percent in blister, and 58 percent in refined. The increase in the share of concentrates reflects the growth of mine output in developing countries, many of which do not have processing facilities or have facilities for only a portion of their output. Important mine copper producers such as Indonesia and PNG have no smelters or refineries, while Mexico and the Philippines have facilities for only a portion of their concentrate output. At the same time there has been a rapid growth in smelting and refining capacity in large consuming countries, notably Japan and West Germany, that import concentrates. The decline in the proportion of blister copper exported is explained in part by the fact that a number of developing countries that formerly exported substantial amounts of blister, such as Chile, Peru and Zaire, have recently installed refining facilities. These developments have given rise to problems of balance between production of concentrates and availability of processing capacities. This has led at times to an excess of concentrates and at other times to an excess of smelting and refining facilities. In 1984-85 there was a shortage of concentrates in relation to smelting capacity, but this condition could change if more U.S. smelters are shut down and a substantial amount of concentrates shipped abroad for processing. On the other hand, some countries, such as Mexico and the Philippines, that have been exporting concentrates are expanding domestic processing facilities.

NOTES

1. By rise in the real exchange value of the dollar I mean a rise in the value of the dollar in terms of other major currencies adjusted for the

change in the U.S. price index relative to the change in the combined price indexes of other major countries.

2. The World Bank includes South Africa and China in developing countries. In this book South Africa is in the developed country category and China is a communist or "centrally planned" country.

REFERENCES

Bureau of Mines (1985b) "Copper," Mineral Facts and Problems, Bulletin 675, U.S. Department of Interior, Washington.

CIPEC (1985) Survey of Mine, Unrefined and Refined Capacities, Paris, June.

Commodities Research Unit (1985) Copper Studies (a monthly publication), New York, October.

Phelps Dodge Corporation (1985) Unpublished tables prepared by the Research Staff, New York, November.

Sassos, Michael P. (1986) "Mining Investment 1986," Engineering and Mining Journal, January 1986.

Siedenburg, William G. (1984) and (1985) Copper Quarterly, Smith Barney, Harris Upham and Company, New York, October 19, 1984 and June 10, 1985.

Takeuchi, Kenji, John E. Strongman, Shunichi Maeda and Suan Tan (1986) The World Copper Industry: Its Changing Structure and Future Prospects, Staff Commodity Working Paper No. 15, World Bank, Washington.

World Bank (1985) Commodity Trade and Price Trends, Johns Hopkins University Press, Baltimore.

Chapter 3

CONSUMPTION AND THE SLOWDOWN IN DEMAND

Copper consumption in the market economies grew at an annual rate of 4.7 percent over the 1950-1973 period, but declined to only 1.2 percent during 1973-1984. There were a number of factors responsible for this decline, some of which had to do with conditions peculiar to the demand for copper while others contributed to the decline in rates of consumption of all major metals, including aluminum, copper, lead, zinc, tin, nickel and steel. Table 3.1 shows the changes in rates of growth in consumption of six major metals for the 1961-1973 and 1973-1984 periods. Much of the decline between the two periods can be explained by the reduction in rates of growth of industrial production and real GDP.

The early post-World War II period was one of rapid industrial growth that accompanied reconstruction and reconversion of the wartime economies. Industrial production grew at an annual rate of 4.7 percent over the 1961-1973 period, but at only 3.0 percent during the 1973-1984 period. There was also a decline in the annual rate of growth in real GDP in the industrial countries from 4.5 to 3.0 percent between the two periods. The greater slowdown in industrial production relative to real GDP for the industrial countries reflected a change in the composition of GDP. This composition in industrial countries has shifted in favor of services, such as education, health and government, relative to industrial commodities that use a higher proportion of metals. Moreover, many qualitative improvements in autos and other consumer durables tend to be less metal intensive in terms of value added, but require more capital and labor to produce. The metals intensity of industrial output expressed in terms of metal tonnage per unit of industrial output (1970=100) for 1982 was 76.9 for nickel, 78.5 for zinc, and 86.4 for copper (Auty, 1985, pp. 275-83). This is a variation of the intensity of use concept employed by W. Malenbaum, which measures the ratio of per capita consumption of raw materials to per capita real GNP (Malenbaum, 1977). Malenbaum found that U.S. copper consumption per billion dollars of GDP (in constant prices) declined by 28 percent between 1951-1955 and 1971-1975, and he projected a nearly 50 percent decline from 1951-1955 to the year 2000 (Malenbaum, 1977, p. 76).

The intensity of use of copper differs with the stage of a country's economic development. Although intensity of use has been declining in industrial countries, it has been rising in developing countries, whose per capita consumption of goods containing substantial amounts of copper has been increasing. Given the higher rates of population growth and the relatively high rates of per capita income growth in developing countries, the rate of growth in copper consumption has been considerably higher than in the developed countries. This may mean that future rates of growth in world consumption will be higher than present rates that are heavily weighted by consumption in the industrial countries. A 1984 World Bank study projected an annual rate of growth of refined copper consumption for industrial countries of 0.9 percent for the period 1985-1995 and a corresponding rate of growth in developing countries of 3.4 percent, yielding a combined annual rate of growth for both groups of 1.5 percent per annum (World Bank, 1984, p. 24). A U.S. BOM study projects an annual rate of growth in U.S. copper demand of 1.9 percent for the 1983-2000 period and of 2.7 percent for world consumption during the same period (Bureau of Mines, 1985, p. 23). Different assumptions regarding key variables, such as rates of growth in GDP, are used in these two projections.

The sharp decreases in rates of growth in consumption between 1961-1973 and 1973-1984 for the six metals listed in Table 3.1 may be explained by changes in the composition of industrial production, the substitution of materials used in the manufacture of end products, and by a variety of technological changes affecting demand for these metals. With respect to composition of output, not only has the rate of growth in the manufacturing sector declined relative to the services sector, but fixed investment in such traditional heavy metal-using industries as steel and railroads has stagnated, while investment in high technology and communications equipment has increased rapidly. The high technology industries have increased consumption of new industrial materials, such as beryllium, tantalum, titanium, lithium and platinum, and a range of nonmetals, such as polymers and optic fibers.

Table 3.1
Rates of Growth in Consumption of Major Metals in Market Economies and Industrial Production and Real GDP, 1961-1973 and 1973-1984 (percent)

	Growth rates[a]	
	1961-73	1973-84
Aluminum, primary	9.1	1.4
Copper, refined	3.9	1.2
Lead, refined	3.1	-0.2
Zinc	4.5	-0.3
Tin	1.4	-2.4
Nickel	7.0	1.3
Steel	5.3	-2.6
Industrial production index[b]	4.7	2.9
Real GDP[b]	4.5	3.0

a Least squares trend growth rates.

b Industrial countries only.

Source: Takeuchi, et.al. (1986) The World Copper Industry: Its Changing Structure and Future Prospects, Staff Commodity Working Paper No. 15, World Bank, Washington, November, p. 41.

Substitution and Conservation

Substitution of other materials has been an important source of reduction in the rate of growth of copper consumption. The degree of such substitution differs greatly among specific uses. Copper has a variety of properties that make it useful either in pure or alloyed form in a number of applications. These properties include electrical and thermal conductivity, resistance to corrosion, strength, and malleability, and physical attractiveness. No other single material has all these characteristics, but several have some that are important in specific uses. The first important substitute for copper was aluminum in overhead high-voltage power cables. There are limitations on the use of aluminum in certain other electrical applications, such as high voltage under-ground cable and in wire used in buildings. Where space is important (as in the case of underground cable) copper is a better conductor than aluminum on a diameter-of-wire basis.

A substantial degree of substitution has also taken place in the communications industry due to development of microwave radios, satellite transmission, fiber optics, and other technological innovations. These substitutions have had a major impact on the use of copper in trunk systems connecting one central office with others. The use of fiber optics is also expected to reduce copper used in urban cables. While fiber optics has important technological advantages for residential use, the installed cost is greater than copper cable (Commodities Research Unit, April 1984, pp. 2-3). Telecommunications wiring accounts for 12 to 15 percent of annual U.S. copper consumption. The BOM estimates that annual consumption in telecommunications lines will be 40 percent lower by the early 1990s, partly because of fiber optics and partly because of the introduction of electronic carrier systems that increase the message-carrying capacity of copper lines. This corresponds to a 5 to 6 percent loss of the total market (Bureau of Mines, 1985, p. 15).

Aluminum wire has been substituted for copper in buildings, but in the U.S. it has been making a comeback due to electrical fires attributed to the use of aluminum. In the postwar period there was widespread use of copper as a substitute for iron, steel and lead in plumbing. But more recently, copper has been losing its market share of plumbing and tubes for heating to plastics. The displacement by other materials has differed considerably from country to country. For example, aluminum radiators in new autos in Western Europe are used in 65 percent of all cars, while in the U.S. and Japan the percentage is substantially lower.

There has been a tendency to economize on all metals used in production because of increased energy costs. A good example is the reduction in use of metal in radiators. Metals technology has produced thinner-walled and smaller-gauged copper tubing for plumbing systems. The revolution in design of electrical and electronic circuits and components has also resulted in a reduction in the use of copper wire. These trends appear likely to continue (Commodities Research Unit, 1982, pp. 1-5).

PRICE SENSITIVITY TO DEMAND

In the above discussion of the factors affecting copper consumption, no mention has been made of price. Since the bulk of consumption is in industrial uses, demand for copper is a derived demand. Moreover, copper inputs constitute a small fraction of the cost and price of autos, electric power, telecommunications services, buildings, and electrical equipment. Therefore, even a substantial change in the price of copper is unlikely to affect demand for these products. Changes in technology have influenced substitution more than changes in price.

There have been a number of attempts to estimate the elasticity of demand for copper in terms of its price relative to the prices of principal substitutes (cross-price elasticity) (Bozdogan and Hartman, 1979, pp. 131-163; and National Materials Advisory Board, 1982a and 1982b). Aside from the fact that investigators' results differ widely, there are serious difficulties in using calculated elasticities as a basis for long-term projections, even if we knew the future course of copper prices.

Substitutes for copper exist for separate uses--aluminum for power transmission, plastics for plumbing and heating, and more recently optic fibers for telecommunications. At any point in time, changes in the price of copper relative to the price of a substitute in a particular application will affect the demand for that application. However, there does not exist a long period during which a given relationship between relative price and demand will hold. Therefore, past relationships cannot be expected to hold in the future. Undoubtedly the copper/aluminum price ratio influenced the rate at which aluminum was substituted in overhead power lines during the early postwar period, but certain technological developments were necessary before that substitution was possible and once made could not be economically reversed, regardless of the relative price of copper. Technological developments in plastics were necessary before they could be substituted for copper or other materials in plumbing and heating, but technology keeps changing with consequent effects on the cost advantage of one material over another. This is also true of the relationship between copper and optic fibers in various uses. For these reasons it is not possible to find a period in the past for which price-demand relationships can be calculated that will hold either in the present or in the future. In forecasting future demand, consideration may be given to projecting relative prices of copper and other material inputs, but this is only one element in calculating the relative cost of using one material input over another. The more important elements in projecting demand are changes in demand for the specific end-products, the capital and operating costs of alternative production processes, and expected new technological developments in the production process.

LONG-RUN DEMAND PROJECTIONS[1]

Long-run projections of the demand for copper are based on simulation models that take into account (1) projections of GNP growth in major consuming countries or classifications of countries by stage of development; (2) projections of the share of industrial production in GNP; (3) the expected development and adoption of technology resulting in materials substitution; and (4) the projected price of copper. The last variable presents difficulties since future prices can only be estimated by modeling both supply and demand using an econometric model that takes into account the relationships among prices and the variables that determine both supply and demand (see Appendix 4-1). It is also necessary to take into account the supply of copper scrap. Scrap represents 25 percent of U.S. refined production, but a lower percentage for the rest of the world. Some scrap is used directly in production without refining so that total copper consumption is much larger than refined consumption. The difficulties in modeling long-run supply have led some investigators to estimate supply changes directly from estimated production from planned and potential projects and depletion of existing mines. Prices are estimated from projected demand-supply balances and adjustment made for the effect of changes in relative prices on materials substitution.

Table 3.2 shows several recent projected rates of growth in copper consumption. The lower growth estimates reflect lower GNP and industrial production growth rates and higher rates of

Table 3.2
Historic and Projected Rates of Growth in World Consumption of Refined Copper (Excluding Soviet Bloc Countries) (percent)

Historic	
1961-1970	4.0
1970-1982	1.9
Fischman[a]	
1980-2000	1.8
World Bank[b]	
1984-1995	1.3
Pollio[c]	
1981-2000	2.1
Bureau of Mines[d]	
1983-2000	2.7

a L. L. Fischman (1980) World Mineral Trends and U.S. Supply Problems, Johns Hopkins University Press, Washington.

b Takeuchi, et.al. (1986) The World Copper Industry: Its Changing Structure and Future Prospects, Staff Commodity Working Paper No. 15, World Bank, Washington, November.

c Gerald Pollio (1983) "The Outlook for Major Metals to the Year 2000: An Updated View," The Journal of Resource Management and Technology, 12:2, April.

d Bureau of Mines (1985), "Copper," Mineral Facts and Problems, Bulletin 675, U.S. Department of Interior, Washington.

substitution of other materials. A World Bank staff study (Takeuchi, et.al., 1986, p. 117) projected an annual rate of growth in consumption for the

1984-1995 period of 1.3 percent.[2] Most earlier projections are 1 to 1.5 percentage points higher. The BOM (1985) projection is 2.7 percent per year for the 1983-2000 period, but in 1980 the BOM projected a rate of growth in consumption for the 1978-2000 period of 3.6 percent per year. Projections by most investigators of the rate of growth in consumption for the rest of the century have tended to decline fairly steadily over the past ten years, and in 1985 were one-third to one-half of the 1975 estimates. Unfortunately, many decisions to invest in increased copper capacity were made on the basis of earlier projections.

Projections of copper demand are highly sensitive to assumptions for the rate of world economic growth. Rates of growth in consumption differ between industrial and developing countries, and also among developed countries or regional groupings, such as the U.S., Japan, and the European Community (EC).

Copper's chief rival in the competitive materials market has been aluminum. Although aluminum has advantages in a number of applications, especially where weight is important, copper has certain qualities that are superior to aluminum. The market for aluminum has benefited from extensive R&D on new applications and their marketing possibilities. The copper industry has spent relatively little on R&D--Alcoa alone spends more on market research and development than the entire world copper industry (Commodities Research Unit, April 1984, p. 15). There are important market opportunities for copper that could be developed. For example, copper roofing has become increasingly popular in Western Europe. Copper on ship hulls as an anti-fouling device has been shown to be well worth the extra cost in terms of saving on fuel and maintenance. Another application is in nuclear waste storage tanks. A decision by the U.S. government to use copper in its nuclear waste storage facilities could create a major market (Commodities Research Unit, 1984, p. 15). However, these possibilities are unlikely to have a major impact on the demand for copper.

The International Copper Research Association sponsored a study by the Technology Assessment Group of Schenectady, New York to examine the effects of new and emerging technological developments on copper demand over the next fifteen to twenty years (Lillie and Abetti, 1986). Those technological areas having a high probability for increasing the demand for copper (in order of their positive impact) include heat pumps, solar power, desalinization, electric vehicles, computers, power electronics, transmission cables, telecommunications, and lasers. Those technological developments posing the greatest threat to copper consumption included fiber optics, aluminum radiators, and reduced use of copper in circuit boards. The overall impact of new technologies on demand was found to be modestly positive. However, I believe this conclusion is too optimistic in view of recent developments in materials technology.

CONCLUSIONS ON THE RATE OF GROWTH IN DEMAND

Given the uncertainties regarding factors determining the long-run demand for copper between now and the end of the century, we should properly consider maximum and minimum rates of growth rather than specific rates. It appears unlikely that the rate of growth in consumption will equal that of the recent past, about 2 percent per year. On the other hand, the annual rate could well be as low as 1 percent. Since a specific growth rate assumption is required for estimating trends in capacity and prices in Chapter 8, I shall assume a rate of 1.5 percent per year.

The abrupt slowdown in demand for copper beginning in the mid-1970s and continuing in the 1980s was mainly responsible for the current over-capacity and low copper prices. Investments in new capacity during the 1970s were made on the expectation that annual consumption would continue to grow at 3.5 percent per year. These investments were also predicated on a rise in real copper prices during the 1980s and 1990s. There was sufficient idled and underutilized copper capacity in 1984 to more than satisfy projected consumption to 1990 and perhaps well beyond that date. Yet there is a substantial amount of planned new capacity, some of which will most certainly be built. Does this mean overcapacity and low copper prices will continue well into the next decade? An attempt will be made to shed light on this question in subsequent chapters.

NOTES

1. A cooperative study by the BOM and the Department of Commerce projected an increase in U.S. copper consumption from 1,664 thousand mt in 1982 to 2,128 thousand mt in 1993, or an annual increase of 2.25 percent. However, 1982 is a poor base year from which to make projections since it was a recession year. If 1972, when U.S. copper consumption was 2,037 thousand mt, had been taken as the base, the average annual rate of growth projected to 1993 would have been less than one-fourth of 1 percent (Bureau of Mines, 1986, p. 59).

2. The World Bank staff projection is a "base case" or "most likely" projection of the rate of growth in world copper consumption between 1984 and 1995. A lower rate of growth in world GNP over the 1984-1995 period than that assumed in the "base case" would result in a somewhat lower rate of growth in world copper consumption, while a higher rate of growth of world GNP would result in a higher rate of growth in copper consumption.

REFERENCES

Auty, Richard (1985) "Materials Intensity of GDP," Resources Policy, 11, December.

Bozdogan, Kirkor and Hartman, Raymond S. (1979) "U.S. Demand for Copper: An Introduction to Theoretical and Econometric Analysis," in Raymond F. Mikesell, The World Copper Industry, Johns Hopkins University Press, Baltimore.

Bureau of Mines (1985) "Copper," Mineral Facts and Problems, Bulletin 675, U.S. Department of Interior, Washington.

_______ (1986) Domestic Consumption Trends, 1972-82 and Forecasts to 1993 for 12 Major Metals, U.S. Department of Interior, Washington, January.

Commodities Research Unit (1984) "Varying Patterns of Copper Consumption," Copper Studies, New York, April.

_______ (1982) "Factors Affecting Long-Term Demand," Copper Studies, December.

Lillie, David N. and Abetti, Pierre A. (1986) "Technological Opportunities and Threats for Copper," presented at the Annual Meeting of the Society of Mining Engineers, New Orleans, LA, March 3-6, 1986. The papers presented were published by the Society of Mining Engineers, Littleton, Colorado.

Malenbaum, W. (1977) World Demand for Raw Materials in 1985 and 2000, National Science Foundation, Philadelphia.

National Materials Advisory Board (1982a) Mineral Demand Modeling, National Academy Press, Washington.

_______ (1982b) Analytical Techniques for Studying Substitution among Materials, National Academy Press, Washington.

Takeuchi, Kenji, John E. Strongman, Shunichi Maeda and Suan Tan (1986) The World Copper Industry: Its Changing Structure and Future Prospects, Staff Commodity Working Paper No. 15, World Bank, Washington.

World Bank (1984) Prospects for Major Primary Commodities: Vol. IV, Washington, September.

Chapter 4

PRICES, COSTS AND THE COMPETITIVE STRUCTURE OF THE INDUSTRY

The world copper industry has been beset by a high degree of price variability, a long-run downward trend in real copper prices, and rising operating and capital costs. The U.S. copper industry has also been adversely affected by the sharp appreciation of the dollar during the 1981-1985 period. U.S. mining companies have responded to declining prices in various ways, such as restricting output, closing mines, selling unprofitable mining assets, and reducing costs by modernizing and reducing wages. Third World mining firms on the other hand have tended to maintain or even increase production in the face of declining metal prices. Mining officials in the U.S. have advocated various forms of government action, such as import barriers, while governments of Third World copper exporting countries have sought international support of schemes for raising commodity prices. These demands for governmental action are examined in Chapter 6. This chapter is concerned with the behavior of prices and costs and how these variables have affected the competitive structure of the copper industry. It is also concerned with how the current demand and supply imbalance in the copper industry is likely to be resolved and the effects of the adjustment process on various producing countries and firms. This analysis will throw light on the question of whether national or world welfare would be improved by some form of public action to stabilize prices, or whether social welfare is maximized by allowing market forces to determine the future structure of the industry.

DEVELOPMENTS IN THE WORLD MARKET

During the 1970s and early 1980s the world price of petroleum was maintained well above the competitive level until 1985-86 when the OPEC cartel lost control over the supply of oil. By contrast, the world market for copper has been quite competitive and for several decades no group of producers has been able to maintain the world price even for a short period of time. Nevertheless, the world copper market does not conform to the economists' concept of perfect competition.[1] First, copper is not a homogeneous product. Not only are different qualities and shapes of refined copper sold on the market, but it is sold in the form of ores, concentrates, blister and various grades of refined metal. Second, contrary to the concept of perfect competition, not all producers sell as much as they can produce at the market price. Some reduce output or accumulate inventory when they cannot sell all they can produce at the price they offer to consumers. In other words, these firms sell at a "producer price" rather than at the free market price. In periods when demand was heavy relative to supply, some firms rationed output among customers rather than raise their producer price by the maximum amount the market would allow. Although there is arbitrage among various grades and forms of copper, arbitrage is not perfect and price differentials exist among markets. Finally, the bulk of the world's copper is marketed under contracts between producers and buyers. Most contracts for various grades of refined, blister, and concentrates specify prices quoted on the London Metal Exchange (LME) or the New York Commodity Exchange (COMEX) as the price to be paid by the buyer to the seller on delivery of the product. In the U.S. and certain other countries, producers of refined copper establish a producer price that may vary from the price of a particular grade on the commodity exchanges. Nevertheless, despite market imperfections and price differentials in world markets, prices tend to move together. Arbitrage prevents the differentials from becoming very large.

Another characteristic of the world copper industry is the limitation on new producers entering the market. Unlike the 19th century when mines could be initiated with a few thousand dollars and there were thousands of independent copper miners, a modern mine costs hundreds of millions or even billions of dollars to construct. This limits entry to large MNCs or government entities. Losses in the industry over the past decade have tended to further reduce the number of firms producing copper so that concentration is increasing.

Yet another characteristic of the world copper market that departs from perfectly competitive conditions is that many government-owned or controlled producers have goals other than profit maximization. A government-owned firm may choose to maintain production and employment even though prices fall below operating costs. In other words, these firms are willing to accept losses in order to maintain employment. Yet another characteristic of the market has been the tendency of some governments to expand the capacity of their copper industries without due consideration of whether prospective profits will justify investment. Even the most careful cash flow analysis has not prevented private investors from making investments in projects that have resulted in losses. This is due to difficulties in forecasting costs and prices for projects that require several years for construction and a decade or more of production before capital costs can be recovered and the expected rate of return on investment realized.

The Producer Price System

Prior to the 1970s, the U.S. producer price system dominated the North American copper market. A few large U.S. and Canadian multinational companies not only supplied 80 percent of the U.S. market, but controlled imports through their ownership of most of the output in Latin America. European copper companies also employed producer prices and largely controlled imports from Africa. U.S. and Canadian producer prices quoted by the major primary copper companies were kept in line by the practice of price leadership, while European producer prices tended to follow LME quotations. However, European producer associations were able to exert considerable influence over the LME price. These conditions changed during the 1970s with the loss of U.S. and European control over production and marketing in developing countries. Competition from imports greatly weakened the producer price system. Some U.S. companies, including Kennecott and Anaconda, began basing their prices on COMEX quotations, while those companies retaining the traditional producer price adjusted prices more frequently with changes in the COMEX. In 1984 Phelps Dodge, the largest U.S. copper producer, began selling wirerod at the average of the COMEX spot price in the month of shipment plus a seven cent per pound premium. However, Phelps Dodge still uses a producer price for cathode and wirebar. Other large U.S. copper producers, including ASARCO and Newmont, continue to sell at the producer price.

Prior to 1978, U.S. and Canadian producer prices were changed infrequently and at times were substantially above or below the LME price, with a differential of as much as 30 cents per pound. Beginning with the sharp rise in the LME price in 1964, the average U.S. producer price was significantly below the LME price through 1970 and in some periods was half the LME price (see Table 4.1). In 1971 the LME price declined substantially; the two prices were approximately the same during 1971-72. In 1973 LME prices were again more than 20 cents per pound higher than the average U.S. producer price, but from 1975 on, the U.S. producer price has been higher than the LME price. It is easy to understand why the U.S. primary copper industry used its market power in the domestic economy to maintain producer prices above outside market prices. It is more difficult to explain why the industry maintained producer prices substantially below the LME and other open market prices for extended periods of time. The reason given by the industry is that a more stable price will induce consumers to use copper rather than turn to substitutes in periods of shortage and relatively high prices. Experience has shown that once consumers shift to substitutes they are unlikely to return to copper. Thus in periods of heavy demand in relation to supply, U.S. producers have rationed their supplies of copper, favoring regular customers with long-term contracts, rather than raise prices in line with those on the LME and COMEX. It is also argued that if consumers obtain lower than free market prices from producers during temporary periods of tight supplies, they will continue to buy at higher prices when free market prices decline in periods of overcapacity.[2] However, consumers are not willing to pay large premiums over open market prices for long periods of time.

Beginning in 1978 not only did some major U.S. copper companies abandon the producer price system, but producer prices themselves were more frequently adjusted with changes on the commodity exchanges,

Table 4.1
Copper Prices, 1960-1985

	U.S. Producer Price[a] (cents /lb)	LME Price[b] (cents /lb)
1960	32.0	30.7
1961	29.9	28.6
1962	30.6	29.6
1963	30.6	29.3
1964	32.0	43.9
1965	35.0	58.7
1966	36.2	69.1
1967	38.2	51.2
1968	41.8	56.1
1969	47.5	66.3
1970	57.7	63.9
1971	51.4	49.3
1972	50.6	48.5
1973	58.9	80.8
1974	76.6	93.1
1975	63.5	56.1
1976	68.8	64.0
1977	65.8	59.4
1978	65.6	61.8
1979	92.3	90.1
1980	101.4	99.3
1981	83.7	79.5
1982	72.9	67.2
1983	77.9	72.2
1984	66.8	62.6
1985	65.6	64.9

a Annual average cash wirebar price.

b Annual average cash settlement wirebar price to 1981; higher-grade cathode thereafter.

Sources: Commodities Research Unit, Copper Studies, New York, May 15, 1981, p. 1 for 1960-1980; CIPEC, Quarterly Review, Paris, January/March 1986, p. 51 for 1981-1985.

and the margins between producer prices and commodity exchange prices declined. Producer prices have been higher than LME prices, but for reasons discussed below, most producers basing their contract prices on the LME and COMEX (both in the U.S. and abroad) charge a premium above the commodity exchange price.

As real prices declined in the 1980s, U.S. mining firms responded by reducing mine output and closing high-cost operations. Anaconda, the third largest domestic producer, closed all its mines; while Kennecott, Phelps Dodge and ASARCO closed some of their large mines and smelters. The reduction in U.S. output was accelerated by the need to meet EPA pollution abatement standards and the price of copper did not warrant large expenditures for replacing or renovating substandard equipment. Mining complexes continuing to operate in the U.S. have lowered their costs by reducing wages and introducing cost-saving technology.

Pricing on the Basis of Exchange Quotations

The abandonment of the producer price system has meant that copper prices are more volatile and prices used in all transactions tend to move together to a greater degree than in the past. Prices are mainly determined in open markets and are less subject to influence by large producers. However, there are considerable differences in contract terms provided by sellers. There are several types of refined copper products (such as wirebar and cathode) and quality classifications for each type that give rise to significant price differences among the products. Producers also brand their standardized products and develop consumer loyalty by assuring a uniform product. How are these market conditions dealt with in the context of a pricing system based on quotations on commodity exchanges?

Transactions on the LME and COMEX constitute a very small percentage of total trade in copper products, but quotations on these exchanges serve as reference prices for most contracts. Most contracts use LME quotations as the reference price, but several U.S. and Canadian producers use the COMEX as the basis for pricing under sales contracts and for contracts to purchase concentrates. CODELCO uses COMEX quotations on contracts for refined copper sold in the U.S. market, but uses LME quotations in contracts for sales to the rest of the world. However, the actual prices charged in contract sales usually differ from quotations on the exchanges for several reasons.[3] First, quotations on commodity exchanges reflect sales of copper products that may differ from the grade and form of products sold under contract outside the exchanges. Only two contracts for refined copper are traded on the LME, one for a standard grade that includes both wirebar and cathode, and the other mainly high-quality brands of cathode and some wirebar. Several types of refined copper are traded on the COMEX, but as yet there is no high-grade contract. Since the actual copper traded under contracts outside a commodity exchange is likely to be of a uniformly higher grade than that quoted on an exchange, these contracts usually provide for a premium of several cents per pound over the quotation used as the reference price. Second, contracts for sales outside a commodity exchange provide for different quotation periods with respect to the reference price, or may give the buyer an option among quotation periods. For example, the quotation period may be the average of the reference price for a month prior to delivery, or for the month prior to the month of shipment, or for the month after shipment. Third, firms purchasing from producers may take delivery wherever they wish, while under LME contracts purchasers must take delivery from one of several warehouses located in Britain and Continental Europe. The conditions outlined above also apply to contracts using a COMEX reference price.

Hedging

Since spot and futures contracts are traded on commodity exchanges, they provide a means by which both buyers and sellers can hedge contracts against changes in market prices over the contract period. A copper fabricator placing an order for future delivery, say, at a price based on the market price at the time of delivery, may want to be protected against an increase in price between the time the contract is negotiated and the time of delivery. If the price rises, he will, of course, gain on the futures contract to buy copper at the current price and this gain will offset the loss from the purchase contract. Likewise, a seller can hedge against a decrease in price between the time a sales contract was negotiated and the time of delivery. Trade in copper futures on the commodity exchanges is actually more important than trade in (spot) contracts for immediate delivery.

The Market for Concentrates

A substantial amount of world mine copper output is produced by firms that do not have smelters or have insufficient capacity for their concentrate output. However, some mining firms have excess smelting capacity and there are firms that specialize in smelting and refining concentrates produced by others. The latter are known as custom smelters and refiners. Japan and Western Europe have much larger smelting and refining capacities than mining capacity, so they import large amounts of concentrates from other countries. Prior to the 1980s, the U.S. had excess smelting and refining capacity and purchased concentrates from Canada, Latin America, and the Philippines. However, since 1981 the U.S. has tended to have a shortage of smelter capacity as a result of shutdowns due to inability to meet environmental standards. Therefore, in recent years a portion of U.S. mine concentrate output has been exported. In 1983 world exports of concentrates totaled nearly 1.5 million mt of recoverable copper.

In most cases custom smelters and refiners buy concentrates under long-term contracts. In the U.S. there is also a certain amount of "toll" smelting and refining. Under "toll" contracts the refined copper and byproduct metals remain the property of the concentrate producer and are returned for subsequent sale after smelting and refining charges are paid. Where concentrates are purchased under a custom contract, there are two elements in the purchase price: (1) the market value of copper contained in the concentrates; and (2) the smelting and refining charges. These charges are deducted from the market value of the copper in calculating payment to be made to the seller of the concentrates. The price under a custom contract is usually tied to the LME or COMEX, or U.S. producer price. The value of byproducts, such as gold and silver, are added to the market value of copper contained in concentrates. Smelting and refining charges may be a fixed percentage of the current price, or a fixed amount per pound, or a fixed amount per pound subject to escalation with changes in certain cost elements. In recent years treatment charges have tended to vary with the supply of concentrates relative to the demand for them by custom smelters and refiners. In 1981 when there was an ample supply of concentrates available in the international market, treatment charges were about

9 cents per pound, but in 1983-84 when concentrates were in short supply, some charges were about half that amount. The decline in treatment charges reflected the sharp reduction in mine output, particularly in the U.S. Low treatment charges reduce the demand for concentrates since low profitability causes custom smelters and refiners to curtail or shut down operations.[4]

The Merchant Market

The merchant market consists of dealers that do not own mining or smelting or refining facilities and who deal in a variety of copper products, including concentrates and scrap. They sell refined copper to fabricators who do not have contracts with primary producers or who want to acquire additional supplies. Merchants may obtain copper from U.S. producers who want to sell a portion of their output at prices lower than their producer price, or from mining firms in various parts of the world that want to sell copper not contracted for sale under long-term arrangements. They may also obtain supplies through commodity exchanges, or they may conduct arbitrage operations in response to price differentials in various markets. The decline in the producer price system has tended to increase activity in the merchant market. Merchants handle about 30 percent of the trade in concentrates and arrange toll contracts for suppliers of concentrates. Prices in the merchant market tend to follow those on the commodity exchanges (Sacks, 1982, pp. 20-22).

The Scrap Market

Since nearly half of total U.S. copper consumption is supplied by scrap, the scrap market is very important (Commodities Research Unit, September 1984, p. 8). High-grade scrap is used with primary copper (from mines) for the production of refined copper; while secondary refined copper is produced entirely from scrap. Scrap is also used directly in production of alloys, such as brass. There are several grades of scrap, the purest being "new" scrap--a byproduct of fabricating operations.

"Old" scrap comes from discarded copper-containing materials and equipment. There are a number of categories of old scrap depending on the composition of the materials. Secondary refined copper is competitive with primary copper in most uses and, therefore, scrap prices tend to fluctuate with prices of refined copper on commodity exchanges. The market is highly competitive because there are thousands of dealers that supply metals for recycling. Hence, in periods of low copper prices, scrap collection and production of secondary refined copper tends to decline, while in periods of high prices they tend to rise. This helps to moderate fluctuations in prices arising from factors affecting the demand and supply for primary copper.

Fluctuating Exchange Rates and Copper Prices

Quotations for copper on the LME are in sterling. While the LME price is widely used as a reference price in contracts, other currencies, such as the dollar or the Deutschemark, are frequently quoted in trade between buyers and sellers. Prices on the COMEX as well as U.S. producer prices tend to fluctuate with the dollar equivalent of the LME price since arbitrage prevents prices on the two exchanges from differing widely. The average LME settlement price for high-grade cathode was £1049 per mt in 1983, while in 1984 the average price was £1032, a decline of less than 2 percent. However, the U.S. dollar equivalent of the LME price was 72.2 cents per pound in 1983 and 62.6 cents in 1984, a decline of 13 percent. There was also a decline of approximately the same percentage in the average price on the COMEX between 1983 and 1984. Between 1982 and 1984 the average LME sterling price rose by 22 percent while the dollar LME price fell by 7 percent. These divergent price movements in terms of the two currencies over the same period raise several questions. The most obvious is whether the world price of copper rose or fell. Under a system of floating exchange rates for which there is no international standard of value, this question cannot be answered.

A more relevant question is how much of the decline in the dollar price of copper was due to a shift in the world demand-supply balance and how much was due to the appreciation of the exchange value of the dollar. Several studies have been made of this problem (Commodities Research Unit, June 1981, pp. 1-6; Radetzki, September 1985). There is considerable evidence that the decline in the average dollar price of copper between 1983 and 1984 was mainly the result of the appreciation of the dollar over this period. The dollar appreciated against a basket of fifteen major currencies by 7.2 percent between 1983 and 1984. In real terms, taking account of relative changes in the price indexes between the U.S. and the countries identified in the same group of currencies, the dollar appreciated by about 5 percent between 1983 and 1984. In 1984 consumption of refined copper in the market economies rose as a result of the world recovery, while production declined by about 1.5 percent between 1983 and 1984. Commercial stocks also declined during 1984. Thus, in the absence of an appreciation of the dollar it might be expected that the dollar price would rise, or at least not decline significantly. A statistical analysis of the various factors affecting the dollar price of copper shows that had the U.S. exchange rate remained the same in 1984 as in 1983, the price would have been about 70 cents per pound in 1984, or some 12 percent higher than the actual average of 62 cents per pound (Takeuchi, et.al., 1986, p. 69).

Since dollar prices of copper declined while dollar prices of other goods and services rose between 1983 and 1984, U.S. producers found their revenues declining and their costs rising. As with producers of other internationally-traded goods, copper producers were the victims of dollar appreciation. However, U.S. fabricators gained from the decline in the dollar price of copper; their copper costs declined while the costs to fabricators in nondollar countries remained about the same. The sharp dollar appreciation over the 1980-1985 period provides, in part, an explanation of why U.S. copper production declined drastically after 1981, while some foreign producers expanded output and production in the aggregate tended to rise.

TRENDS IN PRICES, 1900-1985

Throughout the present century copper prices have fluctuated widely, reaching a peak of about $2.10 per pound (in 1983 dollars) in 1964, after which prices trended downward. Following a sharp rise in prices during World War I, prices declined, rising modestly with the boom period of the late 1920s, but fell to an historic low of about 50 cents per pound (in 1983 prices) during the depression of the 1930s (see Figure 4.1). Despite a decline in average U.S. ore grades from about 2.5 percent at the turn of the century to about 0.5 percent in 1980, productivity rose steadily with technological advances in mining and metallurgy. The number of pounds of copper per man-hour in the U.S. rose from 9.9 in 1912 to about 32 in 1944 (U.S. Federal Trade Commission, 1947, p. 115). Periods of overcapacity followed by rates of growth in consumption exceeding rates of growth in world productive capacity were mainly responsible for sharp price fluctuations. Also, copper was subject to governmental price controls during World War II and during periods of inflation in peacetime--both in the U.S. and Western Europe.

Following World War II and the removal of government price controls, copper prices rose rapidly from an average of about 70 cents per pound (in 1983 dollars) during the war period to more than twice that level in the mid-1950s. The U.S. government stimulated domestic production by subsidies and accumulated copper in the national defense stockpile. The rapid expansion of world copper capacity relative to demand during the 1955-1962 period resulted in a decline in prices until 1963-64 when they rose to their highest level in the present century. During the period of excess capacity (1955-1962), prices were supported by planned cutbacks in production by the major producers that controlled much of the world's output.

During the 1964-1973 period, consumption grew at an average rate of 4.8 percent per year and prices soared, averaging $1.67 per pound (in 1983 dollars). Capacity also expanded, but lagged behind growth in consumption. Also during this period output and capacity growth were impaired by strikes, civil disturbances, and by a number of nationalizations of Third World copper mines owned by multinational mining firms in Chile, Peru, Zaire and Zambia. As was noted in Chapter 3, the rate of growth of copper consumption in industrial countries declined sharply after 1974, and real prices have been trending downward since that time due to overcapacity in the market economies and loss of control of the market by large multinational mining firms. Although sterling prices on the LME firmed somewhat after

Figure 4.1
LME Copper Prices in Constant 1983 Dollars, 1900-1985

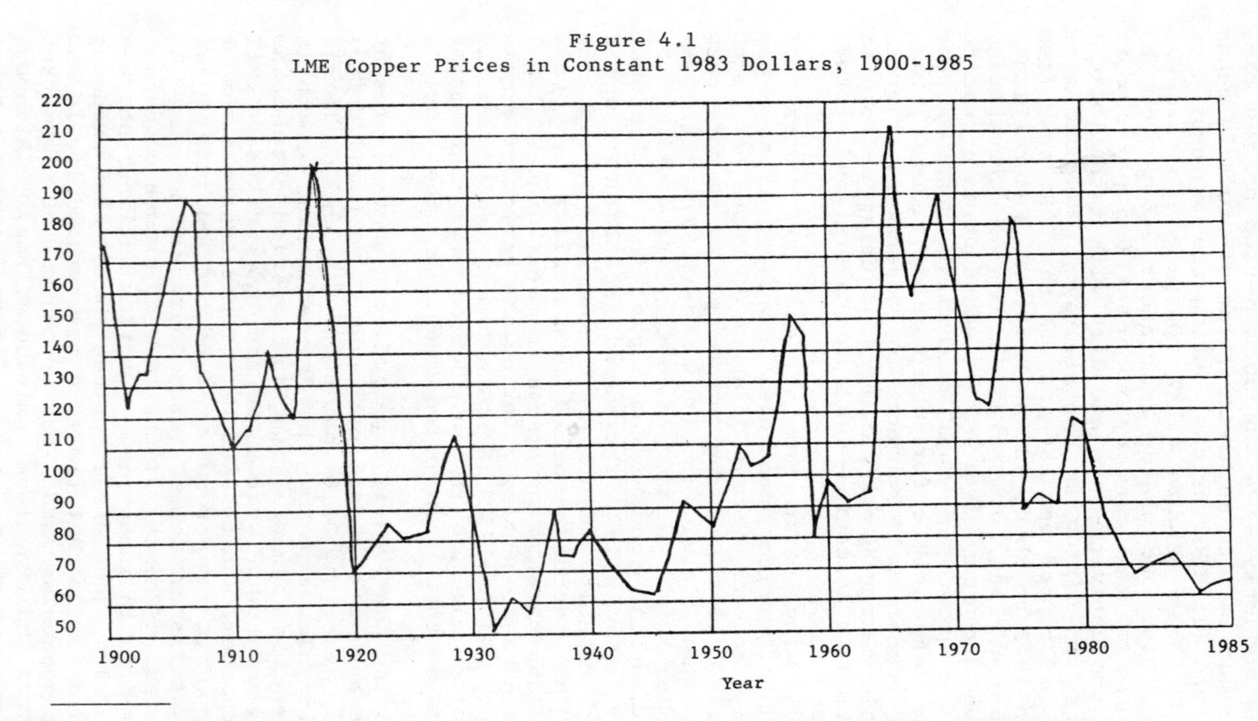

Source: Kenji Takeuchi, John Strongman, Shunichi Maeda, and Suan Tan (1986) *The World Copper Industry: Its Changing Structure and Future Prospects*, Commodity Staff Working Paper No. 15, World Bank, Washington, November, p. 10.

1983, dollar prices continued to decline in 1984 and remained relatively low in 1985 because of the appreciation of the dollar.

Volatility of Prices

Copper prices have fluctuated widely throughout much of the present century. Price volatility is a characteristic of most major metals since they are price inelastic and subject to large shifts in demand over the business cycle, while capacity changes very slowly. Fluctuations in copper prices have been greater than those for most other metals due to the highly competitive market structure and the relatively large number of producers. There is a much greater concentration of production in the hands of a relatively few producers in the case of aluminum, chromium, cobalt, manganese, nickel and tin. Also, most large copper firms outside the U.S. are not integrated into fabrication. The absence of producer control over U.S. copper prices has been especially evident since producer prices began to follow movements on commodity exchanges more closely than was the case prior to the mid-1970s.

Table 4.2 provides a comparison of annual fluctations in prices (in current dollars) of major metals over the 1978-1983 period. Although it is obvious that copper prices fluctuated more than most other metals (with the notable exceptions of lead and molybdenum), the table does not present a complete picture of price volatility. First, information on prices in all markets and the quantities transacted at those prices is not available. For example, merchant market prices for commodities such as chromium, cobalt and nickel have been much more volatile than the producer prices shown in Table 4.2. Second, current price movements do not distinguish between price changes peculiar to the market structure of particular metals and those arising from world inflation. The 1978-1983 period was marked by a relatively high rate of increase in industrial prices, the composite index of which rose by 45 percent over the period. Price stability for a metal in constant prices would exist if the annual increase in its current price corresponded to the rise in the general index of prices. For none of the metals in Table 4.2 has the average annual price moved proportionately with the U.S. index of industrial prices. In fact, prices for some metals,

Table 4.2
Average Annual Prices of Major Metals, 1978-1983
(U.S. dollars)

Commodity	1978	1979	1980	1981	1982	1983
Aluminum (in pounds)						
Producer price, ingot	.54	.61	.72	.76	.76	.78
Bauxite (dollars per mt)						
Guyana	138	153	212	216	208	180
Chromium (per mt chromite)						
South African producer price	6-20	20-25	25	17-25	12-17	12
Copper (range in dollars per pound)						
U.S. producer wirebar	.67	.93	1.02	.85	.73	.78
LME	.62	.90	.99	.79	.67	.72
Iron Ore (dollars per mt)						
Brazil (North Sea ports)	19.39	23.44	27.24	24.62	26.21	23.97
Manganese Ore (dollars per long ton)						
U.S. ports	144	140	155	168	164	152
Nickel (dollars per pound)						
Canadian ports	2.09	2.71	2.96	2.70	2.19	2.12
Lead (in pounds)						
U.S.	.34	.53	.44	.37	.27	.23
LME	.30	.55	.41	.33	.25	.19

Table 4.2 (cont.)

Commodity	1978	1979	1980	1981	1982	1983
Molybdenum (in pounds)						
U.S. price in concentrates	4.95	7.50	9.70	8.50	4.00	3.64
Tin (in pounds)						
Penang market	5.68	6.72	7.46	6.38	5.87	5.91
Zinc (in pounds)						
U.S. high grade metal	.32	.38	.38	.46	.40	.43
LME	.27	.34	.35	.38	.34	.35
U.S. Index of Producer Prices	100	111	126	138	143	145

Sources: Bureau of Mines (1984) _Mineral Commodity Summaries 1983_, U.S. Department of Interior, Washington; and International Monetary Fund (1985) _International Financial Statistics Yearbook 1985_, Washington, 1985.

including copper, declined after 1980, while the U.S. index of industrial prices rose sharply between 1980 and 1983.

One measure of the degree of fluctuation in commodity prices is the average percentage deviation of real prices from a moving average over a given time period. Table 4.3 shows the average percentage deviation from a five-year moving average (in constant 1981 dollars) for nine major metals over the 1955-1981 period. Copper has the highest degree of real price fluctuation followed by lead and zinc. This may be explained in part by the low degree of concentration of production in these commodities, and by the fact that all three are traded on one or more commodity exchanges. Until recently, nickel prices were subject to substantial control by Canadian producers, and tin prices were controlled by the International Tin Agreement.[5] The production of chromium, cobalt and manganese is also subject to a high degree of concentration and this facilitates price control by producers.

Table 4.3
Average Percentage Deviation of
Real Prices from Five-Year Moving Average,
1955-1981

Copper	15.3
Lead	14.7
Zinc	14.4
Tin	8.1
Manganese	7.2
Iron Ore	5.7
Bauxite	5.6
Nickel	4.6
Aluminum	4.6

Source: World Bank (1983) The Outlook for Primary Commodities, Washington, p. 45.

TRENDS IN PRODUCTION COSTS

Economic theory teaches that over long periods of time the cost of producing a metal, including discovery, capital and operating costs plus a rate of profit sufficient to justify new investment, determines its real price. In a competitive industry with free entry, a period during which prices remain substantially higher than full production costs will induce investment in new capacity, while a long period during which real prices are below costs will retard new investment and marginal mines will be shut down. However, because considerable time is required before new investment can bring additional capacity on stream and because mines can operate for long periods when prices do not cover full production costs, prices fluctuate well above or below full production costs. Moreover, there are dynamic factors continually affecting costs so that the marginal cost of additional output from some new mines may be well below the average cost for the industry.

Comparisons of product costs over time are useful for explaining long-run price trends, and cost comparisons among countries suggest changes in comparative advantage likely to affect the future distribution of output. However, comparisons of production costs present difficult conceptual problems. First, full economic costs should include some costs not accounted for in company reports, such as the return on the equity investor's capital (including an allowance for risk), that would warrant a similar investment today. In corporate reports interest on indebtedness is included in costs, but not after-tax returns on equity investment. Hence, mining firms with large indebtedness would show a large cost for the use of capital while a firm of similar size with no indebtedness would show none. Also, depreciation and depletion are determined by tax regulations rather than actual depreciation and depletion. Second, replacement costs for a mine built twenty-five years ago may be several times the original cost. To be useful, cost comparisons between countries and over time must refer to full economic costs for specific periods, with adjustments for changes in the industrial price index. Third, the cost per pound for copper shown in company statements is net of the value of byproducts. For mines with high gold and silver values, the net per pound cost of copper may be very low and, in any case, costs will vary with the market values of gold and silver. Thus, when the price of gold rose sharply in 1979-1981, some relatively high-cost copper mines suddenly became very low-cost and highly profitable. Fourth, costs among countries are sensitive to changes in exchange rates since it is necessary to convert costs in domestic currencies into a common currency--usually the dollar. In the case of developing countries, the official exchange rate is frequently overvalued, thereby tending to artificially raise domestic costs when converted into dollars. The 40 percent appreciation of the dollar between 1980 and 1985 in relation to other major currencies had the effect of raising U.S. production costs relative to those of other countries when their domestic currency costs were converted into dollars. Hence, 1983 or 1984 estimates of U.S. production costs may be regarded as inflated relative to costs in the second half of 1986 when the exchange value of the dollar was substantially lower. Finally, production costs among countries are greatly affected by national tax systems. Royalties and other taxes not based on profits are regarded as cost elements in production, but profits taxes are not. However, profits taxes, which range from 40 to 70 percent, greatly affect the return on investor capital. In Peru, high-grade ores and low labor costs make for relatively low operating costs, but mining companies regard Peru as a high-cost country because of its high corporate income tax. With all these difficulties in mind, we may now look at some comparisons of average copper production costs over time and among producing countries.

Historical Comparisons of Production Costs

Orris Herfindahl (1959, Chs. 4,5, and 6) found long-run costs of producing copper remained fairly steady between 1885 and World War I, after eliminating abnormal years.[6] However, because of technological advances in mining and processing ore, production costs declined by about 40 percent during the second decade of this century and remained fairly stable between the early 1920s and 1957 (Herfindahl, 1959, Ch. 2 and pp. 58-59). Several opposing developments combined to maintain the real cost of copper during the period between World War I and 1957. First, there was a decline in the grade of ore mined. In the U.S. the average grade mined was 1.6 percent in 1921, and 0.8 percent in 1951-1956. Opposing the declining U.S. ore grades were (a) technological advances and scale economies that tended to reduce real costs per pound; and (b) the rapid development of mines in Latin America and

Africa where ore grades are much higher and labor costs lower.[7]

A study undertaken by the CRU (March 1983, p. 18) showed that while U.S. ore grades fell by one third between the mid-1950s and the late 1960s, labor productivity in terms of pounds of copper per man hour rose by 60 percent. However, there were substantial increases in wages during that period. CRU estimates world production costs in constant 1981 dollars averaged 50 to 55 cents per pound in the 1950s, rose to 60-65 cents in the late 1960s, and were about 64 cents in 1973. Thereafter, costs declined to 52-53 cents in 1979-80 mainly due to the increase in byproduct prices. However, the fall in precious metals prices in 1981 resulted in an increase in average production costs to around 72 cents per pound. These estimates were based on costs in major producing areas in the market economies, with North American costs averaging 77 cents as contrasted with 69 cents in Latin America, 80 cents in Africa, and 90 cents in Europe.

In summary, the historical record indicates that average real production costs of copper declined between the late nineteenth and the early twentieth century. They remained fairly stable from the early 1920s until the 1960s when they rose from an average of 50-55 cents per pound in the 1950s to about 64 cents in 1973 and to about 72 cents in 1981 (all in 1981 dollars). Production costs rose in the second half of the present century, mainly as a consequence of the increase in petroleum prices and in capital costs for new copper producing capacity. The latter included an increase in the cost of mining equipment, a rise in real interest rates, and a relative shift from equity to debt financing.

Comparing Costs Among Countries

Comparing copper production costs among countries yields information regarding their relative competitive advantage. However, future investment and output in a country may depend on factors unrelated to costs, such as the level of taxation, general investment climate, and the organizational structure of the industry. For example, a mining industry largely controlled by an SME may not be affected by the tax structure and may not have the same investment objectives as a country dominated by private investment. There are usually wide differences in costs of mines within a country that are masked by country averages. In fact, a country such as the United States with relatively high average costs has mines whose costs are well below the world average. Also, future costs may be as affected by development of new mines with ore grades substantially higher or lower than the country average. Introduction of new technology as well as other cost-reducing innovations may affect a country's average costs. Changes in byproduct prices result in significant changes in relative costs for countries with ores containing substantial byproducts. Finally, relative production costs are constantly influenced by changes in real exchange rates (nominal rates adjusted for inflation relative to inflation abroad) and by changes in real rates of interest for mine financing. Thus rates of interest on loans to SMEs are usually lower than those paid by private mining companies.

Despite these complications, several studies have been made of the relative average costs of producing copper in major producing countries. Such studies generally include the following:

1. Direct costs--mining and milling, smelting, refining, transporting and marketing costs;
2. Indirect costs--administration, research, exploration, royalties and other taxes, except corporate income or profits taxes;
3. Depreciation of structures and equipment;[8]
4. Interest on loans; and
5. Revenue from byproducts and coproducts deducted from costs listed above to yield net costs.

Most cost estimates do not include a return on equity capital or allow for the replacement cost of the mine. Hence, the estimates do not reflect the full cost of production.

A World Bank staff study (Takeuchi, et.al., 1986) showed wide differences in average net costs per pound (gross costs less co/byproduct revenue) among twelve major producing countries (see Table 4.4). (In this study interest on debt was included, but not depreciation.) The relative cost positions of individual countries shifted substantially from 1975 to 1980 and from 1980 to 1984, the three years for which production cost data were estimated. According to the study, the three countries with the lowest average costs in 1975 were PNG, Mexico and Canada, in that order; the highest cost countries were the U.S., Zambia and Zaire. In 1980 the lowest cost producers were Canada, PNG and Australia; the highest cost producers were Zambia, the U.S. and the Philippines. In 1984 the lowest cost producers were

Table 4.4
Production Costs of Major Copper Producing Countries
1975, 1980 and 1984
(cents/pound)

Country	Gross Costs			Co/Byproduct Credits			Net Costs		
	1984	1980	1975	1984	1980	1975	1984	1980	1975
U.S.	88.7	105.9	68.6	10.6	32.5	7.0	78.1	73.4	61.6
Zambia	74.8	93.4	63.2	7.8	9.1	1.6	67.0	84.0	61.6
Australia	103.9	78.6	51.7	37.6	50.9	38.3	66.3	27.7	38.3
Peru	112.2	129.9	111.6	55.4	88.8	60.5	56.8	41.1	51.1
Canada	182.2	158.5	118.7	126.2	168.1	90.3	56.0	-9.6	28.4
Philippines	92.8	101.4	49.8	37.3	44.1	11.7	55.5	57.3	38.1
Chile	59.8	76.9	52.7	11.1	20.1	5.5	48.7	56.8	47.2
Indonesia	65.6	63.9	46.2	19.6	30.6	10.7	46.0	33.3	35.5
South Africa	78.7	98.7	48.8	33.1	56.0	7.5	45.6	42.7	41.3
Zaire	69.8	120.2	67.8	24.6	69.1	12.7	45.2	51.1	55.1
Mexico	104.5	139.1	87.0	71.6	97.0	87.0	37.9	42.1	27.3
PNG	88.9	109.1	48.3	56.6	91.2	24.5	32.4	17.9	23.8
Average[a]	93.4	108.7	70.9	36.9	58.8	22.0	56.7	49.9	48.9

a Includes Sweden.

NOTES: Gross costs include (a) direct operating costs of mining and milling, smelting and refining, transporting, marketing costs; (b) indirect costs include administration and corporate overhead, research, exploration, royalties, "front-end" taxes (excluding corporate income and profits taxes, and interest on debt. Co/byproduct credits constitute revenues from co/byproducts. Net costs equal gross costs less co/byproduct credits.

Source: Takeuchi, et.al., (1986) The World Copper Industry: Its Changing Structure and Future Prospects, Staff Commodity Working Paper No. 15, World Bank, Washington, p. 60.

PNG, Mexico and Zaire; the highest cost producers were the U.S., Zambia and Australia. In large measure changes in the market value of co/byproduct credits were responsible for shifts in the relative cost positions among the countries studied. Chile and Indonesia had the lowest gross costs in all three periods, but were not among the three lowest net cost producers because their co/byproduct credits were relatively low compared to PNG, Mexico, Canada, Australia, Peru, the Philippines, South Africa and Zaire. The U.S. and Zambia had relatively high gross costs, and relatively low co/byproduct credits. However, gross costs in the U.S. were considerably lower than those for Mexico, Canada, Peru and Australia. Average gross costs in the U.S. declined substantially between 1980 and 1984, while net costs rose mainly due to a decline in gold, silver and molybdenum prices--the major byproducts of U.S. production. The decline in U.S. gross costs was in considerable measure a consequence of the closure of high-cost mines, and increased productivity in the mines that continued to operate. The average gross costs for all countries declined from 109 cents per pound in 1980 to 93 cents in 1984, but net costs rose from 50 cents to 57 cents as a result of a fall in market prices for co/byproducts.

Other cost studies using different methods and cost elements show somewhat different results. For example, a BOM study (Bureau of Mines, 1985, p. 16) shows average cash production costs for copper in the U.S. to be 65 cents per pound in 1984, substantially lower than the figure in Table 4.4.

The changes in the rankings of copper producing countries between 1980 and 1984 shown in Table 4.4 were influenced by changes in the exchange values of their currencies in terms of the U.S. dollar over this period. There was a large appreciation of the dollar against the major industrial country currencies and a much larger appreciation of the dollar against the currencies of some copper producing developing countries. Since the currencies of copper producing developing countries depreciated against the dollar by more than their inflation rates relative to the U.S. inflation rate, the real (or inflation-adjusted) exchange value of their currencies in terms of the dollar declined. Nevertheless, the extent of the real depreciation of the currencies of copper producing countries in terms of the dollar differed significantly among countries. This affected the relative cost rankings among the countries listed in Table 4.4.[9]

In 1985 there was a substantial reduction in production costs in Canada, Chile and the U.S. These reductions are mainly attributed to increased productivity and, in the case of the U.S., to lower wages in the mining industry. Average net costs declined by 13 cents per pound in the U.S.; by 14 cents in Canada; and by nearly 7 cents in Chile. In 1985 average net cost was lower in the U.S. (65 cents per pound) than in Mexico and the Philippines; while Canada, Chile and Peru all had net costs in the 41-42 cents range. On a gross cost basis (excluding co/byproduct credits), U.S. costs of 72 cents per pound were at about the mid-point of the major producers, while Chile's costs (51 cents) were lowest (Takeuchi, et.al., 1986, p. 77).

TRENDS IN REAL COSTS

There is considerable evidence that the real cost of mining copper (adjusted for inflation) increased during the 1970s. Fuel costs rose by a factor of ten between 1970 and 1981. Real wages have risen in nearly all producing countries, but in many countries output per man hour has more than offset the increase in real wages. Environmental costs were higher, especially in the U.S. copper industry. Unfortunately, there is no series of comparable annual data on average real costs of production for the period 1970 to 1985. Copper Studies (Commodities Research Unit, 1980, p. 1) estimates that U.S. average operating costs rose by 9 percent (in constant dollars) between 1970 and 1975, and there were increases in real costs in some but not all other producing areas.

The cost estimates shown in Table 4.4 for 1975, 1980 and 1984 are in current U.S. dollars. For measuring changes in real costs of production it is better to use gross costs since net costs are affected by volatile co/byproduct prices. If we adjust average gross costs by the U.S. producer price index (1975=100), real costs per pound in 1975 dollars increased from 71 cents in 1975 to 72 cents in 1980 and declined to 52 cents in 1984. Thus the average real cost of producing copper was almost constant between 1975 and 1980 and declined by 28 percent between 1980 and 1984. In the U.S. the real cost declined by 29 percent between 1980 and 1984. This was a remarkable achievement considering the sharp rise in the price of energy and, in some countries at least, the necessity of meeting pollution abatement standards. Several factors may be cited for the reduction in both real and nominal costs during the 1980-1984 period. First, low copper prices induced cutbacks or shut downs of high-cost operations, especially evident in the U.S. Second, there was some modernization and introduction of new technology despite the low level of investment in the industry. High-cost smelters were abandoned in favor of selling concentrates to firms with more efficient smelters or in some cases new smelters were built. But perhaps the most important cost reduction occurred with better management, which made possible reductions in the workforce and in overhead costs. U.S. management also succeeded in reducing wages in the copper industry. There were some actions that resulted in only temporary cost reductions, such as the deferment of maintenance and development, reduction in waste stripping ratios, and highgrading of ore. Production was expanded by the treatment of tailings by leaching and electrowinning technology. This process produces copper without any mining costs, but, of course, could not continue indefinitely.

FULL ECONOMIC COSTS OF PRODUCTION

The cost data shown in Table 4.4 does not represent full economic costs of production since it does not include the capital recovery of a mining complex built at today's prices or the return on capital sufficient to attract new investment in copper mining. An important source of increase in total costs of production since 1970 has been the rise in capital costs (Radetzki, 1979, pp. 350-59). Sir Ronald Prain (1975, p. 187) estimated the 1970 average capital cost of producing a metric ton of refined copper at $3,000. A World Bank staff study (1983, p. 92) estimated the average capital cost per metric ton of copper in the early 1980s to be $9,200 (in 1981 dollars).[10] This figure includes capital costs for mining, smelting and refining. In constant 1970 dollars this would be about $3,800, or an increase of some 27 percent above the estimate given by Prain. In addition to capital costs, financing costs have been boosted by a substantial increase in real interest rates in the 1980s over those in the early 1970s, and by a sharp increase in debt-equity ratios for new mining ventures.

The U.S. BOM has estimated the full cost of producing copper in 110 mines operating in the market economies as of January 1985. Full costs include recovery of capital and a 15 percent (after tax) IRR on all capital invested, both equity and debt.[11] The average full production cost for the 110 mines in the survey was 77 cents, well above the 65-cent average price of copper in 1985. The 1985 price would have covered full production costs for only the nine Chilean mines, but was well below the average cost of 82 cents for the thirteen U.S. mines (see Table 4.5). The full production costs for thirteen temporarily shut down U.S. mines was estimated at $1.16.

The above findings do not necessarily mean that most mines now in production will be shut down unless the price of copper rises sufficiently to cover the full production costs of the 110 mines operating in January 1985. Mines can operate as long as revenues cover operating costs and, in the case of some SMEs, government subsidies make up deficits in operating revenues. However, as was noted earlier in the chapter, several hundred thousand mtpy of U.S. copper producing capacity was shut down between 1981 and 1984, and more may be closed in the future unless there is a significant increase in the price of copper. Also, mines require continual investment to maintain a given level of output. In the absence of a price sufficient to warrant this investment, output from existing mines will tend to decline.

The BOM estimated full production costs for ninety-eight explored but undeveloped copper mines in the market economies as of January 1985. Again production costs include capital recovery and a 15 percent (after-tax) IRR on invested capital. The average full cost per pound of refined copper that could potentially be produced by these mines was estimated at $1.67 per pound. The seven Chilean mines covered in the study have the lowest average estimated cost, $1.20 per pound; while the average cost for the thirty-four U.S. explored deposits was $1.94.[12] Average cost for Australian and Canadian explored mines was estimated to be $2.42 and $2.13, respectively. The study suggests that very few new mines will be developed during the next decade or so unless the price of copper rises to at least $1.20 per pound and for the thirty-four U.S. explored

Table 4.5
Estimated Full Production Costs for 110 Producing Copper Mines in the Market Economy Countries, January 1985[a]
(U.S. dollars per pound of refined copper)

Country	No. of mines	Mine oper. cost	Mill oper. cost	Smelter refine cost[b]	(Less) byprod. credit	Net oper. cost[c]	Recov. of capital
Australia	3	.19	.10	.32	.07	.55	.08
Canada	14	.27	.35	.45	.32	.76	.12
Chile	9	.30	.17	.11	.05	.53	.05
India	5	.34	.21	.42	.03	.93	.12
Peru	6	.19	.20	.32	.05	.65	.14
Philippines	7	.31	.27	.22	.21	.58	.07
South Africa	4	.41	.25	.50	.49	.67	.13
U.S.	13	.23	.20	.23	.09	.56	.12
Zaire	7	.39	.19	.25	.41	.42	.07
Zambia	8	.31	.14	.05	.09	.42	.12
Others	34	.25	.27.	.29	.30	.51	.10
Total or Average	110	.28	.21	.22	.16	.55	.09

Table 4.5 (cont.)

Country	Taxes @ 0% ROR[d]	Prod. cost @ 0% ROR[c]	Taxes @ 15% ROR[d]	Ret. on invest. @ 15% ROR	Prod. cost @ 15% ROR[c]
Australia	.01	.64	.06	.07	.76
Canada	.01	.89	.07	.06	1.01
Chile	.02	.60	.04	.03	.65
India	.05	1.10	.20	.11	1.36
Peru	.02	.81	.08	.07	.94
Philippines	.05	.70	.07	.04	.76
South Africa	.00	.80	.04	.04	.88
U.S.	.02	.70	.06	.08	.82
Zaire	.08	.57	.15	.05	.69
Zambia	.06	.60	.14	.04	.72
Others	.05	.66	.11	.09	.81
Total or average	.03	.67	.08	.05	.77

a All production costs estimated in January 1985 U.S. dollars.

b Includes smelting and refining charges, transportation costs to the smelter and refinery (but not to market), and post mill processing charges for other (noncopper) commodities.

c Data may not add to totals shown because of independent rounding.

d Includes property, severance, state and federal taxes and royalties.

Source: Bureau of Mines (1986), (unpublished table).

deposits, the price would need to rise to nearly $2 per pound (in January 1985 dollars).

PRICE OUTLOOK TO 1995

The findings of the 1986 BOM study summarized in Table 4.5 suggest that in the absence of a significant rise in the real price of copper over the next several years, output in the market economies will decline and little, if any, new producing capacity will be developed. These findings imply that after the current overcapacity in the world copper industry has been eliminated by growth in consumption and the permanent closing of mines regarded as temporarily shut down, copper prices must rise to at least $2 per pound in January 1985 dollars sometime in the late 1990s. A problem with this conclusion arises because this level of copper prices is substantially above current projections for the mid-1990s made by the World Bank and other investigators. A World Bank staff study (Takeuchi, et.al., 1986, p. 113) forecasts copper prices to rise from 1986 levels to only 74-cent per pound in 1995 (in 1984 dollars). (This is a base case forecast, but the price of copper might be lower if world economic growth is lower than assumed.) The 74 cent projected price for 1995 is substantially lower than the price the 1986 BOM study estimated would be required to cover the average cost of developing new mines. However, because of existing overcapacity few new mines will need to be built to satisfy the demand for copper in the market economies to 1995. It may be noted there are several large deposits in Chile capable of producing several hundred thousand mtpy that are likely to be developed at a price well under $1 per pound (in 1985 dollars).[13]

A study by Philip Crowson (1983, pp. 38-47) suggests that an average copper price of $1.05 per pound in 1982 dollars ($1.10 per pound in 1985 dollars) is a reasonable projection over the next decade or so. Crowson based his conclusions on a study of cash break-even costs (all cash expenses of producing copper plus interest on outstanding debt less byproduct credits) between 1969 and 1982. He finds that during the period 1971-1980 there was substantial stability in cash break-even costs for the mines representing 85 percent of the average cost curve, and that average annual costs tend to rise and fall with the LME price. Thus, following a sharp fall in LME prices in 1974-75, average cash break-even costs declined as a consequence of the elimination of high-cost production and increased productivity. He states that "prices influence costs almost as much as costs affect prices." (Crowson, 1983, p. 45). Operating costs for some U.S. mines declined by 20 cents per pound when low prices induced cost reductions by the elimination of high cost operations, modernization and renegotiation of wage contracts. However, Crowson's analysis does not take into account the total cost of developing new mines that will eventually be needed to meet demand.

OUTLOOK FOR COMPETITIVE STRUCTURE AND TRADE PATTERNS

The future competitive structure of the world copper market depends heavily on the rate of growth in copper consumption. If consumption to the year 2000 were to grow at the rate of 2.7 percent per year as projected by the BOM, primary copper capacity would need to rise by 35-40 percent between 1984 and 2000, after allowance for existing excess capacity. Under these conditions the real price might rise to $2 per pound as implied in the 1986 BOM study. On the other hand, if between 1984 and 1995 annual consumption grows at 1.3 percent and mine production at 0.8 percent, as projected by the World Bank staff report (Takeuchi, et.al., 1986, p. 113), real prices are likely to rise only modestly and required capacity would be less than 10 percent above the 1984 level. Under the slow consumption growth assumption and with little increase in real prices, U.S. primary copper output is projected to decline from 1.1 million mtpy in 1984 to 975,000 mtpy in 1990 with a further decline to 890,000 mtpy by 1995 (Takeuchi, et.al., 1986, p. 129). Capacity and output in low-cost countries, such as Chile, Mexico and PNG, are likely to rise by more than enough to offset a decline in output in the high-cost countries. Under a high consumption growth rate assumption, some of the shut down U.S. capacity might be restored to production, while most of the growth in output would be provided by low- and medium-cost countries. (Canada, Australia, Philippines, Zaire and Peru fall into the medium-cost group.)

As was discussed in Chapter 3, the low consumption growth scenario appears more likely in the light of current developments. However, several questions arise in forecasting the competitive structure of the world copper market even if the rate of growth in consumption could be forecast with confidence. One is whether a substantial amount of U.S. shut down capacity will be reopened or whether more mines will close during the next few years. Even if prices rose only modestly from the 1985 level of 65 cents per pound to 75 cents per pound, it is estimated that 250-300,000 mt of idled capacity would be restored to production (Demler, 1985, p. 53). Modernization and economies achieved through better management and lower wages could further reduce cash break-even costs for U.S. mines. A second source of uncertainty is the outlook for new investment in a number of developing countries with relatively high-grade undeveloped deposits. The growth of output in Chile, which is not only the largest market economy copper producer but has the lowest costs among major producers, depends upon whether foreign private firms decide to make substantial investments in Chilean mining. CODELCO's output is scheduled to increase over the next few years, but over the long run output from its four large producing mines will decline as a consequence of lowering ore grades. PNG also has large partially explored deposits, but development will require very large capital outlays by foreign investors. The same is true of Peru where the climate for foreign investment is currently very poor and that country's external credit situation is so unfavorable that foreign loans may not be available for large investments by Peru's SMEs. Zaire has very large copper producing potential, but expansion of production will require long-term loans from development assistance agencies and a marked improvement in investment climate. Neither large foreign private loans to the governments nor foreign direct investment in mining in these countries are likely in the foreseeable future. The financing of Third World copper development is discussed in Chapter 5.

PROJECTING CONSUMPTION, PRODUCTIVE CAPACITY AND PRICES

Throughout this study reference is made to projections by various investigators of copper consumption, productive capacities and prices. While in most cases values of these three variables are estimated separately, it is important to recognize that they are mutually interdependent and their values are simultaneously determined at any point in time. In addition, they are affected by a large number of other variables, some of which are exogenous while others react to changes in the values of the three copper variables. In order to understand the complex relationships among these variables and to forecast consumption, productive capacity and prices, econometric models have been formulated by independent scholars, university research institutes, government agencies, international lending agencies and private research organizations that provide consulting services to private companies and public agencies. These models are used to forecast copper prices based on assumed exogenous variables, such as real GNP and its composition, and to determine the effects of changes in specific variables, such as consumption, on other variables, such as prices or capacity. A brief description of econometric models and their uses is given as an appendix to this chapter.

APPENDIX 4-1: ECONOMETRIC COPPER MODELS

Econometric copper models enable us to investigate a number of economic variables and behavioral functions that together determine future copper prices. Behavioral equations are estimated from historical data and applied to estimating future values by using exogenous variables such as forecasts of real GNP. Price movements are determined by shifts in demand and supply with demand and supply reconciled by changes in inventories, but both demand and supply are influenced by prices with time lags of various lengths. Therefore, econometric models must determine a number of endogenous variables simultaneously.

Refined copper is the type usually modeled, but refined output is derived from both mine copper and scrap. Some scrap is used in the production of primary copper, but secondary refined copper is derived wholly from scrap. In addition, scrap is used directly in the production of some fabricated products such as brass, and this demand must be taken into account in estimating the supply of scrap for refined copper. Equations for each source of copper must be estimated separately. Mine output is a function of mine capacity and of capacity utilization. Mine capacity may be increased by the expansion of existing mines and by the construction of new mines. In the former case, two or three years are required, but new mines require five to eight years for construction. The capacity of existing mines declines over time with depletion of the ore body and mines are sometimes permanently closed due to adverse competitive conditions. Capacity changes are a function of price or expected price, and are also a function of productivity which may be increased by new technology or better management. Higher prices may stimulate investment in new capacity or expansion of existing capacity, but there are substantial time lags before production is increased. However, higher capacity utilization usually occurs shortly after a price increase. Scrap collection and processing is also a function of the price of refined copper, again with a time lag.

Demand for copper is a derived demand and therefore depends upon changes in the demand for manufactures in which copper is used. It is not sufficient to estimate demand simply by forecasting aggregate industrial production since changes are continually taking place in the composition of production. Consumption functions must be disaggregated by major end uses for copper. Changes in prices in relation to prices of substitutes also influence consumption. Consumption functions must allow for lags since time is required before changes in price can induce changes in inputs and in the equipment required for an alteration of inputs.

Since copper demand and supply are both determinants of prices and determined by prices, how is it possible to forecast prices? Various theoretical approaches and econometric techniques are used to deal with the problem of simultaneous determination. In a competitive equilibrium, the demand for and supply of a commodity must always be equal, including an allowance for changes in inventory or stocks. The difference in the amount of copper supplied by producers and the amount demanded by consumers must be instantaneously adjusted by a change in inventory. Hence, in most models price depends upon changes in inventories since it is necessary to induce holders of inventories to increase or decrease them when there is an imbalance between producer supply and consumer demand. The explanation of how inventory holders react to changes in price differs among investigators. According to the CRU model, the level of copper stocks in relation to the expected volume of world consumption determines the price of copper. If inventories fall below a certain percentage of expected consumer demand, the price will rise; if stocks rise above a certain percentage of expected demand, the price will fall.[14] This is called the "stock" approach to price determination. In the "flow" approach, the actual price depends on inventory changes and previous prices. Unfortunately, information is available only on portions of the inventory, mainly those held by metal exchanges, by producers in major countries, and by government stockpiles.

Uses of Econometric Copper Models

Investors in the copper industry often base their decisions on forecasts of prices made with the aid of econometric models. The basic models may be adjusted to take into account a number of special assumptions regarding expected changes in technology in both the production of copper and the industries in which copper is used. The models may also be used to evaluate investment risk arising from business recessions or changes in the rate of growth of consumption. By means of simulations, the effects on prices of potential developments may be determined. Fabricators and manufacturers using copper employ forecasts of prices for estimating costs and for deciding among alternative materials. Governments of developing countries whose foreign exchange income is heavily dependent on copper exports may use forecasts of prices to guide them in fiscal and investment planning. Financial institutions base their project loan evaluations on projected prices.

Econometric models have also been used to examine the effects of alternative government policies on the industry. For example, in a study prepared for the EPA Arthur D. Little Inc. (1978) projected the economic impact of environmental regulations on U.S. output of refined copper. Several econometric simulation studies have analyzed the feasibility of an international copper price stabilization program using a buffer stock or quota restrictions (Charles River Associates, 1977). Finally, the CRU copper model was used to forecast the effects on world copper prices of the elimination of exports from certain countries (Commodities Research Unit June 1981, p. 6). Similar studies might be used as a guide to the setting of stockpile goals for various minerals held in the U.S. national defense stockpile.

Econometric copper models have generally been used to forecast prices five to ten years or more in the future rather than for short term forecasts. This is true in part because movements of key exogenous variables, such as industrial production or prices of substitutes, require considerable time to exert their full effects on consumption, production and price. In general, econometric models for forecasting short-run price movements, whether in commodities or foreign exchange, do not perform any better in prediction than forward market prices.

NOTES

1. For a market to be perfectly competitive, the following conditions must hold: (a) there are a large number of firms each with a very small share of the market; (b) the firms produce a homogeneous product using identical production processes and possess perfect information; (c) there is free entry into the industry in the sense that new firms will enter if they observe that greater than normal profits are being earned; and (d) all firms are "price takers" and sell at the prevailing market price as much as they are capable of producing at prices below marginal costs.

2. For a discussion of the rationale for the two-price system, see Mikesell, 1979, pp. 111-16.

3. For a discussion of pricing terms under copper contracts offered by major producers see Commodities Research Unit, December 1984, pp. 1-8 and Pino, July/September 1985, pp. 11-19.

4. For a discussion of the concentrate market, see Commodities Research Unit, June 1984, pp. 1-10 and Hobson, October/December 1984, pp. 40-69.

5. In 1985 the buffer stock manager operating under the International Tin Agreement ran out of funds and lost control of the tin market. Canadian

nickel producers were no longer able to control the price of nickel after 1983. Transactions in nickel now tend to follow LME and COMEX prices.

6. A study by Commodities Research Unit, (January 1983, pp. 10-14) estimated that costs of producing copper fell from $1.36 per pound in 1822 to 80 cents per pound in 1900 (in 1982 dollars). Thereafter there was a further decline in production costs for porphyry copper mines from 71 cents in 1909 to 53 cents in 1920 (in 1982 dollars).

7. Herfindahl had little information on direct production costs, but based his estimates largely on real prices of copper which he assumed reflected real costs.

8. Some cost estimates exclude depreciation. Depreciation and depletion allowances reflect the nature of the tax system and may not indicate actual depreciation.

9. The changes in the relative competitive positions of countries resulting from changes in exchange rates between 1980 and 1984 were also influenced by the relative share of the local portion of production costs. For a discussion of the effects of exchange rate changes on the relative competitive position of producing countries see Takeuchi, et.al., 1986, pp. 68-74.

10. The 1981 capital cost per annual mt of refined copper, including exploration, mine and equipment, and infrastructure for a surface mine was estimated in the 1983 BOM study (Rosenkranz, et.al., p. 21) at $5,800, but to this amount it is necessary to add at least $3,000 for smelting and refining equipment.

11. The initial BOM study published in 1983 (Rosenkranz, et.al., 1983) covered 150 producing mines as of January 1981. The analysis in the text is based on an unpublished preliminary table prepared by the BOM, which updates copper production costs through January 1985 for 110 mines in market economies.

12. The original BOM study based on a survey of 121 explored but undeveloped deposits in the market economies was conducted on the basis of January 1981 data (Rosenkranz, et.al., 1983). The information in the text is taken from a preliminary unpublished table prepared by the BOM in 1986.

13. Planned new mine copper capacity in Chile includes La Escondida (280,000 mtpy) owned by Broken Hill Proprietary (BHP), RTZ, and a Japanese consortium; Cerro Colorado (65,000 mtpy) owned by Rio Algom, a Canadian subsidiary of RTZ; and an

100,000 mtpy expansion of EXXON's La Disputada mine. A substantial expansion is also planned by CODELCO. All these expansions are likely to take place if the price of copper rises to 85 cents per pound. (This conclusion is based on information obtained by the author during a 1986 visit to Chile.)

14. Commodities Research Unit, June 1981, pp. 1-6. For a discussion of other copper models, see Wagenhals (1985, pp. 30-53); Vogely (1975); and Richard (1978).

REFERENCES

Barsotti, Aldo F. and Rodney D. Rosenkranz (1983) "Estimated Costs for the Recovery of Copper from Demonstrated Resources in Market Economy Countries," Natural Resources Forum, 7:2, April.

Bureau of Mines (1985) "Copper," Mineral Facts and Problems 1985, U.S. Department of Interior, Washington.

Charles River Associates (1977) Feasibility of Copper Price Stabilization Using a Buffer Stock and Supply Restrictions from 1953-1976, UNCTAD, Geneva.

Commodities Research Unit (1984) Copper Studies, "1985 Pricing Trends," New York, December.

________ (1984a) "Concentrate Market Roundup," June.

________ (1984b) "Secondary Smelters in the U.S.," September.

________ (1983) "Production Cost History: Part I," January.

________ (1983) "Production Cost History: III," March.

________ (1981) "The CRU Long-Term Copper Model," June.

________ (1981) "The Copper Price and Floating Currencies," December.

________ (1980) "US Copper Production Costs," June.

Crowson, Phillip C.F. (1983) "Aspects of Copper Supply: Past and Future," Quarterly Review, CIPEC, Paris, January/March.

Demler, Frederick R. (1985) "Copper Market Outlook," Metals, Drexel Burnham and Lambert, New York, May.

Herfindahl, Orris C. (1959) Copper Costs and Prices: 1870-1957, Johns Hopkins University Press for Resources for the Future, Baltimore.

Hobson, Simon (1984) "The Current and Future Market for Custom Copper Concentrates," Quarterly Review, CIPEC, Paris, October/December.

Little, Arthur D. Inc. (1978) Economic Impact of Environmental Regulations of the United States Copper Industry (Report Prepared for the U.S. Environmental Protection Agency), Boston.

Mikesell, Raymond F. (1979) The World Copper Industry, Johns Hopkins University Press for Resources for the Future, Baltimore.

Pino, Victor (1985) "The Trade and the Reference Price--The Producer's View," Quarterly Review, CIPEC, Paris, July/September.

Prain, Sir Ronald (1975) The Anatomy of an Industry, Mining Journal Books, London.

Radetzki, Marian (1979) "The Rising Costs of Base Materials--The Case of Copper," Mining Magazine, April.

__________ (1985) "Effects of a Dollar Appreciation on Dollar Prices in International Commodity Markets," Resources Policy, September.

Richard, D. (1978) "Dynamic Model of the World Copper Industry," Staff Papers, International Monetary Fund, Washington, 25, 779-833.

Rosenkranz, R. D., E. H. Boyle, Jr., and K. E. Porter (1983) Copper Availability--Market Economy Countries: A Minerals Availability Program Appraisal, BOM Information Circular 8930, U.S. Department of Interior, Washington.

Sacks, Harold (1982) "Copper's Metamorphosis from a Merchant's Viewpoint," Quarterly Review, CIPEC, Paris, January/March.

Takeuchi, Kenji, John E. Strongman, Shunichi Maeda and Suan Tan 1986) The World Copper Industry: Its Changing Structure and Future Prospects, Staff Commodity Working Paper No. 15, World Bank, Washington, November.

U.S. Embassy (1985) "Chile Industrial Outlook Report: Minerals," mimeo, Santiago, June.

U.S. Federal Trade Commission (1947) "The Copper Industry in the United States and International Copper Cartels," Report on the Copper Industry: Part I, USGPO, Washington.

Vogely, William A. (ed.) (1975) Mineral Materials Modeling: A State of the Art Review, Resources for the Future, Washington.

Wagenhals, Gerhard (1985) "Econometric Copper Market Models," Quarterly Review, CIPEC, Paris, January/March.

World Bank (1986) "Half-Yearly Revision of Commodity Price Forecasts and Quarterly Review of Commodity Markets for December 1985," (mimeo), Washington.

——— (1983) <u>The Outlook for Primary Commodities</u>, Staff Commodity Working Paper No. 9, Washington.

Chapter 5

THIRD WORLD COPPER INDUSTRIES: DEVELOPMENT AND OUTLOOK

INTRODUCTION

In 1981 copper mine capacity in developing countries was 53 percent of the total capacity of the market economies. This percentage rose to nearly 56 percent by 1984 and appears likely to continue to rise throughout the remainder of this century. In 1981, 62 percent of total mine capacity in the developing countries was held by mining firms with majority government ownership. This percentage has remained fairly stable since the mid-1970s. There have been increases in private investment in copper mining capacity in Chile, Mexico, Peru, the Philippines, PNG and Indonesia; while major expansions in capacity have been made by SMEs in Chile, Peru, Iran and Zaire. No important expropriations have taken place in developing countries since 1974. Except in Chile, Mexico and the Philippines, most of the private equity in mine capacity in developing countries is owned by large MNCs.

Since there are very large copper resources and a number of explored but undeveloped deposits in developing countries, the future growth of mine capacity in these countries will depend mainly on conditions for new foreign investment and on the ability of SMEs to obtain foreign capital to finance new investment. The major purpose of this chapter is to assess the outlook for both foreign private and government investment in Third World copper mining. I shall also examine the production and marketing policies of SMEs.

RESERVES AND PRODUCTION COSTS

About 70 percent of the copper reserves and 65 percent of the reserve base in the market economies is in developing countries (see Table 1.1). Moreover, the ore grades of producing mines are substantially higher in developing countries. For example, the average ore grade (in percent) for recoverable copper from explored surface deposits is 0.84 for Chile; 0.87 for Peru; 0.52 for the Philippines; 5.70 for Zaire; and 0.47 for the U.S. For underground deposits the average grade is 0.95 for the U.S.; 0.88 for Canada; 1.20 for Chile; 3.20 for Peru; and 4.08 for Zambia (Rosenkranz, et.al., 1983, p. 7). As was noted in Chapter 4, production costs are considerably lower in major Third World producing countries (e.g., Chile, Zaire, PNG and Peru) than in the U.S., and in some cases lower than those in Canada and Australia. This advantage for developing countries does not necessarily assure their producing capacity will grow faster than in developed countries. With capital costs for mines averaging well over $6,000 per mt (in 1985 dollars), developing countries would need to spend about $6 billion to expand their capacities in line with the projected increases to 1990 shown in Table 2.5.

TRENDS IN FOREIGN DIRECT INVESTMENT (FDI)

After World War II there was a substantial increase in FDI in the mining industries of developing countries, nearly half of which was in copper, with the major investments made in Latin America. Global data on investments of MNCs in developing countries by industry groups are scarce and unreliable. However, since U.S. MNCs account for well over half of the FDI in mining and smelting in developing countries, we may use U.S. data as a proxy for all MNCs. The book value of U.S. FDI in mining and smelting rose from about $750 million in 1950 to nearly $2.5 billion in 1968. This represented an increase of more than twice the book value of such investments in 1950 dollars. However, the book value of FDI declined sharply after 1968 as a consequence of expropriations of U.S. and European investments (mainly copper investments) in Chile, Peru, Zaire, Zambia and other countries. In 1981 the value of investments of U.S. MNCs in mining and smelting in developing countries (in constant dollars) was less than half that for 1970. Between

1981 and 1984 the value (in constant dollars) of U.S. FDI in mining and smelting in developing countries increased by less than six-tenths of 1 percent, and relatively little of this was in copper (Survey of Current Business, 1985, 1986).

Capital expenditures in developing countries by majority-owned foreign affiliates of U.S. mining companies have also declined drastically since the mid-1970s. In 1974 and 1975 these expenditures averaged about $600 million (in 1982 dollars), declining to an annual average of $340 million in 1980 and 1981, and during the 1982-1984 period averaged just over $100 million (Survey of Current Business, 1986). Most capital expenditures in Third World mining during the 1980s have been in mines producing gold, silver and coal.

There has also been a decline in the book value (in constant dollars) of U.S. mining investment in developed countries, particularly in Australia, Canada, and South Africa, between the first half of the 1970s and the early 1980s. Capital expenditures by majority-owned foreign affiliates of U.S. companies have also declined in these countries. However, the decline in U.S. FDI in mining in developed countries has been less than that in the developing countries.

What these data indicate is that FDI in mining slowed to a trickle during the late 1970s and early 1980s. Moreover, with the notable exception of CODELCO in Chile, SMEs in developing countries have not made substantial additions to copper producing capacity since the mid-1970s.

TRENDS IN MINING CONTRACTS

The major reasons for the decline in FDI in developing country copper mining over the past decade were low prices, overcapacity and poor market outlook. In addition, there has been a deterioration in the investment climate brought about by the wave of expropriations in Third World producing countries that occurred from the late 1960s to the mid-1970s. While there is less danger of expropriation and most countries are welcoming foreign investment in their mining industries, the conditions for FDI represented in mining contracts, fiscal systems, and general economic and political environment in developing countries are much less favorable than during the 1950s and early 1960s.

There is a large literature on mining agreements between private investors and host governments and no attempt will be made to provide a comprehensive treatment of this subject.[1] Most modern mining agreements are based on legislation that authorizes the negotiation of agreements by government authorities. Some governments have evolved "model contracts" based on both the underlying legislation and provisions adopted by administrative authorities, often patterned on earlier contracts. In some countries contracts must be submitted to the legislature for final approval, while in others this is not required unless the contract departs in some way from existing legislation. The agreements are closely tied to the fiscal legislation of the host country and, in some cases, special tax arrangements apply to mining. The agreements usually cover several minerals since not enough may be known about the mineralization to determine which minerals can be commercially produced. However, mining agreements usually exclude petroleum, coal, uranium and other radioactive minerals, since such minerals may be exploited only under special arrangements. Most agreements have fixed termination dates, usually no more than thirty years after the beginning of operations, and provide maximum periods for exploration and construction prior to initiation of production. The exploration area may be quite sizeable, in some cases several hundred square miles, the bulk of which must be relinquished during the exploration period so that the actual mining lease may cover no more than ten square miles. An important provision in all agreements is that the contractor has the right to develop the area he has explored. However, a mining lease is generally not provided until after the feasibility study has been completed and the mining company submitted an investment plan for approval by the government. The major features of typical agreements negotiated during the past two decades are summarized in the following paragraphs.

1. The government exercises considerable supervision and control over the investment plan, including the level of expenditures; the capacity of mine and processing facilities; construction of infrastructure, including roads, ports, utilities; arrangement for disposal of overburden and other waste; control of pollution; and manpower training and minimum percentages of domestic employment for each classification--skilled labor, clerical, technical and supervisory, and managerial and

professional--and the time periods during which these percentages must be achieved. The government also establishes regulations regarding procurement of supplies and equipment, with preference given to domestic resources. A good example of government control over an investment plan is provided in the agreement for the development of the Ok Tedi mine in PNG. In the development of this mine, the first three years of operations were to be devoted to mining and processing the gold cap. Both copper and gold were to be mined and processed during the following three years, and thereafter the major product was to be copper. The company failed to provide facilities for copper processing by the end of three years due to low prices and difficulty in obtaining financing, and the government shut down the mine until a new agreement on this phase was reached.

In the 1969 Cuajone agreement between Southern Peru Copper Corporation and the Government of Peru the company was required to submit annual plans and minimum investment expenditures to the government. The agreement stated that failure to meet minimum investment expenditures in any year could result in cancellation of the agreement. The Peruvian government also established a marketing agency which was given a monopoly on all foreign sales of minerals.

Most agreements provide for government control over the export proceeds of the minerals, but there usually are provisions to assure that foreign creditors financing the mine have first claim on the export proceeds.

2. Most modern agreements provide for government equity participation in the mining venture. Governments desire equity participation both as a means of sharing in the profits and of exercising control through the appointment of representatives on the board of directors. In a few cases, such as the Selebi-Phikwe nickel/copper mine in Botswana, the government share is provided without charge. Generally the government is given an option to buy a specified percentage of the shares of the mining company either at the time of the decision to construct the mine or after production begins. For example, the Ok Tedi agreement gave the PNG government an option to acquire up to 20 percent of the shares following the decision to proceed with the project. In the PT Rio Tinto Indonesia agreement (which was never implemented), the foreign investor was required each

year to offer shares to Indonesians (government or private nationals) equal to not less than 5 percent of the total number of shares issued until 51 percent of the total had been offered to Indonesians. In other cases a government may be in a joint venture with a foreign investor. For example, in the case of the Mamut copper mine in Malaysia, the government holds 49 percent of the equity with the remainder held by the foreign investor. In Zambia foreign investment in mining is not permitted unless the government holds 51 percent of the equity; while in Mexico a majority of the equity shares must be held by Mexican citizens and management must be under Mexican control. Foreign investors generally reject such conditions.

Where mining contracts provide an option to the government to acquire a certain percentage of the shares, the price paid is usually based on book value or the total capital expenditures made by the private investor. This has the disadvantage of diluting the private investor's equity since book value is usually substantially less than the true value of the equity shares. The investor has taken the risks during the exploration and development phases and has financed the project for several years before the expiration of the option. In the case of a joint venture agreement, the government usually puts up its share of capital at the time mine construction begins. Thus, the government avoids the risks incurred by the private investor during the exploration period before a decision is made to construct the project.

Despite these disadvantages, mining executives sometimes favor equity participation by the host government. Such participation gives the government a vested interest in the financial outcome of the project and not simply an interest in earning revenue from taxation. Government equity participation may prove useful in providing public officials with a better understanding of the operating problems of the mining enterprise. Government representation on the board of directors provides an avenue for educating public officials regarding problems that may arise between the company and the government.

3. Fiscal arrangements in modern mining agreements tend to emphasize income taxation over other forms of taxation. Prior to World War II royalties and land rents constituted the major source of government revenue from mining. Royalties

represent a charge for the depletion of mineral resources, which in most developing countries are legally owned by the state. Income tax regimes, including so-called excess profits taxes, are aimed more at taxing the economic rents of resource exploitation. Economic rent is the excess of revenue from a project over the full cost of production, including a "normal" return on equity investment. Typically the fiscal arrangement provides for a normal corporate tax applicable to all industries; rapid depreciation of all capital expenditures; and some form of excess profits tax. Where royalties are assessed they tend to be low, ranging from 1 to 3 percent.

Provision is often made for accelerated depreciation or amortization. Up to three-fourths of the construction cost of most large mines is provided by foreign loans and credits with maturities ranging from six to eight years following initiation of production. Repayment of such loans usually requires all or most of the net operating profits during the first few years of operation, so that little is left for taxes. Rapid capital recovery assures that no corporate taxes will be levied until all or most of the loans have been repaid.

4. Normal corporate profits taxes in developing countries range from 40 to 50 percent, but most governments levy excess profits taxes in order to capture a larger share of any economic rent. There are two general types of excess profits tax: (a) a percentage tax on income in excess of a specified accounting rate of return based on accumulated capital expenditures; and (b) a percentage tax on the accumulated present value of net cash receipts at a given rate of accumulation. Under the first type, profits in excess of the amount that would raise the ratio of accounting profits to invested capital over a 15 or 20 percent threshhold in any given year may be taxed at 60 or 70 percent. For example, the Indonesian model mining agreement provides for a 60 percent tax on profits in excess of a 15 percent rate of return on funds invested. Investors dislike this type of tax since their profits tend to fluctuate from year to year, and in a year of exceptionally high profits they may be paying a marginal rate of 60 to 70 percent, while average profits as a percentage of invested capital over a ten-year period may be less than 10 percent.

Under the second type of excess profits tax, the tax is not applied until the investor has earned a minimum IRR on all capital outlays up to the year the tax is levied. This method is best represented by PNG's "additional profits tax" embodied in the Ok Tedi mining agreement. Under this fiscal system the investor recovers all his capital outlays at a discounted cash flow (DCF) rate of 20 percent (or 10 percent plus the rate of interest on U.S. AAA bonds, at the option of the investor); any amounts above the cash flow necessary to achieve this DCF rate are taxed at the rate of 70 percent. This approach, which has been called the resource rent tax (RRT) is illustrated in Appendix 5.1 of this chapter. The RRT avoids the problems associated with fluctuating accounting rates of return by assuring the investor a minimum IRR before the excess profits tax is levied.

Despite the advantages of the RRT over other types of excess profit taxes, it has certain disadvantages from the standpoint of both the foreign investor and the host government. First, in applying the RRT to foreign investors whose home governments allow foreign tax credits against tax obligations arising from dividends transferred to the home country, such tax credits would not be available until the RRT was actually paid. In the case of a 38 percent accounting profits tax, the entire amount could be used to offset the investor's U.S. tax obligations. The amount of the RRT in any one year may be larger than the amount that can be credited against the tax obligation of the foreign investor to his home government in that year. For example, if the foreign investor's home government taxes accounting profits transferred in any year at 38 percent while the RRT is 60 or 70 percent of the accounting profits for that year, only a portion of the tax paid to the host government can be credited against the tax obligation to the home government.

A second objection to the RRT, or any excess profits tax, is that when companies invest in high-risk projects, they expect to be able to earn a very high return on the successful ones. Hence, the RRT tends to discriminate against high-risk projects with a low probability of success.

A third objection is that an investor might be earning fairly high accounting returns for a number of years before the RRT cuts in. Hence, there may be political opposition to the arrangement within the host country if the government receives no revenue prior to the time the RRT is applied. This situation could lead to a demand for contract revision. In practice, the third problem has been dealt with by the imposition of a tax on accounting profits in addition to the RRT.

Chilean Mining Contracts

A notable exception to the mining contract terms described in the preceding paragraphs is found in the mining agreements in Chile. Chilean mining law does not require comprehensive controls over investment plans and project operations. There is no requirement that the Chilean government participate in ownership of foreign or domestic mining companies, although the government has negotiated joint venture agreements with foreign investors when there is a mutual advantage in doing so. Mining leases may be held in perpetuity and be transferred to other investors without government permission. There is no excess profits tax and mining is subject to the same corporate tax rate that applies to all other industries. Finally, Chile's foreign exchange regulations with respect to transfers of profits and capital are more liberal than most other developing countries. However, there is political and economic instability in Chile and this presents the risk to potential investors that contracts may be changed by some future government.

Contract Violation

Although government expropriation of mining properties has been rare since the mid-1970s, there is considerable risk of contract violation. In part this risk arises from economic and political instability in some of the major copper producing countries of Africa and Latin America. Economic instability together with inability of countries to service foreign debt has increased risk for both foreign equity investors and suppliers of credit. Political instability also increases the likelihood that leftist governments will come to power and violate existing contracts or demand renegotiation. The unstable conditions in many Latin American and African countries provide little confidence that over the next decade or two existing governments will not be displaced by governments that refuse to honor agreements negotiated by their predecessors. In the case of Peru, the present government has violated petroleum contracts previously negotiated.

The outlook is better for Asia-Pacific countries, such as Indonesia, Malaysia and PNG, but it should be noted that during the past decade these governments have demanded renegotiation of some contracts with mining and petroleum companies.

The risk associated with foreign investment in developing countries is heightened by two legal principles relating to contracts with private firms. One is that a contract may be altered by the government whenever conditions change (or are alleged to have changed) from those existing at the time of negotiation. This legal principle, referred to as *rebus sic stantibus*, is invoked when a government finds it convenient to demand renegotiation of a contract with a foreign investor. The other legal principle adopted by virtually all Latin American countries is that disputes over contracts involving a government and a foreign investor must be settled by a national court and cannot be submitted to international arbitration. Experience has shown that national courts rarely find in favor of foreign investors and against governments. Adherence to this principle explains why no major Latin American country is a member of the International Centre for Settlement of Investment Disputes (ICSID) sponsored by the World Bank. However, a number of Asian and African developing countries are members. In some cases, submission of a dispute to ICSID for arbitration is explicitly provided in mining contracts.

OUTLOOK FOR FOREIGN INVESTMENT IN MINING

Assuming a rise in world copper prices to 75 or 80 cents per pound (in 1986 dollars), what is the outlook for FDI in copper mining in Third World countries? For most copper producing countries the outlook is exceedingly poor over the next decade. Although the bulk of Peru's copper output is produced by an American firm (SPCC), Peru has a history of mine expropriation and contract violation, and foreign mining companies are subject to the highest mining taxes in the world. In addition, Peru is experiencing high inflation and political unrest, and the government has defaulted on its foreign debt. India, the Philippines, Mexico and Zambia have not permitted majority-owned foreign investment in their mining industries in recent years. A variety of adverse investment conditions exist in Zaire. The only foreign-owned copper mine,

Sodimiza (owned by a Japanese consortium), was sold to the Zaire government in 1983. Societe Mineriere de Tenke-Fungurume (SMTF), created in 1970 by a consortium of American and European firms to produce 150,000 mtpy of copper, was liquidated in 1984 without reaching the production stage--at a loss to investors of over $200 million. SMTF was unable to complete its mine because the railroad route through Angola was closed and bank loans arranged for the project were cancelled. Brazil has copper deposits, but that country's mining industry is dominated by the SME, CVRD, and there is little foreign investment, except on a joint venture basis.

Among the developing countries, only in Argentina, Chile, Indonesia and PNG are conditions favorable for FDI in copper mining. There are relatively rich deposits in Argentina, Chile and PNG that have been explored by foreign investors and some are likely to be developed pending an improvement in the price. Political and economic conditions in PNG are relatively stable and additional FDI is likely during the next decade. Provided a moderate democratic government is established in Chile before the end of the decade, that country is also likely to receive a substantial amount of additional FDI. Indonesia welcomes foreign investment, but until recently its mining contracts have not been attractive to investors and there are no immediate prospects for investment in copper. Argentina's copper production is currently negligible, but it has substantial potential; a project for producing 100,000 mtpy is planned by a U.S. MNC. However, the financing for this project, estimated to cost $1.2 billion, is unlikely to be available unless there is an improvement in Argentina's external debt position.

Altogether, new copper capacity developed by foreign investors is unlikely to exceed 530,000 mt over the next ten years. The potential expansion includes 100,000 mt in Argentina, 350,000 mt in Chile, 100,000 mt in PNG, and 80,000 mt in Peru.

POLICIES OF STATE MINING ENTERPRISES

Three decades ago there were almost no state-owned mining enterprises in developing countries, but by 1981, 62 percent of mine copper capacity, 73 percent of smelter capacity, and 77 percent of refining capacity in these countries was produced by firms with majority government ownership

(Radetzki, 1985, p. 20). In Brazil, Zaire and Zambia 100 percent of the output is produced by SMEs; in Chile, 90 percent; India, 65 percent; and Peru, 10 percent. In addition, there was significant government ownership of mining enterprises in developing countries producing 73 percent of mine copper capacity, 76 percent of smelting capacity and 83 percent of refining capacity. The bulk of the mining capacity that is majority-owned by government enterprises was originally developed by MNCs, so that governments have been responsible for developing less than 25 percent of the existing mining capacity in developing countries. In Mexico there is mixed domestic private and government ownership of the mining industry, but the major mines were discovered and developed by U.S. companies. There has been substantial expansion of Mexico's copper producing capacity in recent years.[2]

SMEs differ considerably in technical capacity, experience and managerial talent.[3] Some, including Chile's CODELCO and Peru's Centromin, inherited a fully trained and experienced managerial and technical staff when the mines were nationalized. After operating in the more advanced countries for several decades, foreign-owned enterprises usually reduce their expatriate staff to only a handful. However, in more primitive countries, such as Zaire and Zambia, the nationalized mines continue to employ thousands of expatriates in managerial, technical and professional positions. In both of these countries a substantial portion of the expatriate staff stayed on after nationalization and there were management contracts between the nationalized companies and the former owners. Since these contracts were terminated, the SMEs in Zambia and Zaire have had difficulty in maintaining sufficient expatriate staffs and in training domestic staffs to fill vacancies. These difficulties are reflected in the low productivity of their mines compared with very high productivity of CODELCO's mines.

In general, the technical capacity of SMEs is good and the best compare favorably with the best operated mines owned by MNCs. They are also quite capable of exploring deposits and contracting new mines. In fact, both SMEs and MNCs negotiate contracts with the same large international construction and engineering firms, such as Bechtel Corporation, Arthur C. McGee, or Parsons-Jurden.

But technical competence alone does not assure good investment decision-making or the appointment of competent managers. Investment decisions are sometimes influenced by noneconomic factors such as the desire to promote regional development or to increase employment rather than to maximize returns on capital.[4] In addition, management positions are sometimes filled with political appointees.

The policies and financial performance of SMEs vary widely and most generalizations regarding how they differ from those of private companies are wrong or misleading. Some SMEs maintain output in the face of prices below operating costs, but private mining companies also continue producing under these conditions when closure costs are high. (Closure costs include severance pay, cost of cancelling contracts, maintenance during closure, and start-up costs) (Evans, 1986). For some SMEs the major constraint on expansion appears to be their ability to obtain capital rather than whether they can earn a rate of return on additional investment comparable to what might be earned in other industries. This may constitute a misallocation of resources by the central government. However, it is not difficult to find cases of unwise investment decisions by MNCs. There is evidence that SMEs are less efficient and less profitable than private enterprises in similar circumstances (Radetzki, 1985, pp. 40-41). There is also evidence of political influence in the appointment of SME managers. SMEs have social objectives, such as promoting employment and regional development, which frequently compete with minimizing costs and maximizing profits. By contrast, MNCs are primarily profit motivated. MNCs are faced with the financial constraints of borrowing in private capital markets and providing a satisfactory return on stockholders' equity--constraints that do not exist for SMEs that obtain their capital either directly from the government or by borrowing with a government guarantee. Nevertheless, some SMEs earn substantial profits on which they pay the normal corporate tax and, in addition, provide most of their funds for investment from reinvested profits.

FINANCING FOREIGN MINING PROJECTS

Large, modern open pit copper mining complexes, including concentrator, smelter and infrastructure, cost from several hundred million to one billion dollars and require several years to construct. Few MNCs are able to provide all the financing for a project of this size from internal sources, or would want to incur the full risk of such a large investment in a developing country. Therefore, MNCs usually arrange for the bulk of the construction costs to be financed by nonrecourse loans and credits from international banking consortia, equipment suppliers, and government export-financing institutions. Exploration and the feasibility study are financed with equity. In recent years, equity investment has tended to be made by a consortium of companies, with one company designated as manager. The Cuajone copper mine in Peru is owned by SPCC (a consortium of ASARCO, Phelps Dodge, Newmont Mining and the Marmon Group) and Billiton B.V. (a subsidiary of Royal Dutch Shell Petroleum Company). Thirty-six percent of the total cost of the mine ($726 million) was provided by equity and the remainder by long-term debt financing--including $253 million from commercial bank loans, $140 million from equipment suppliers, $54 million from consumers, and $15 million from the International Finance Corporation (IFC) (Mikesell, 1983, p. 115). Foreign equity in the Ok Tedi mine in PNG is held by a consortium of Broken Hill Pty. (BHP), 37.5 percent; Amoco Minerals, 37.5 percent; and a group of German companies led by Metallgesellschaft, 25.0 percent. In the case of the Freeport Indonesia copper mine, the American investor, Freeport Minerals, provided only $24 million in equity or 16 percent of the $146 million capital cost of the first stage of the project. The loan financing was provided by U.S. insurance companies and commercial banks, a German bank, and a group of Japanese copper consumers and trading companies (Mikesell, 1983, p. 137). Loans from U.S. insurance companies were guaranteed by the Overseas Private Investment Corporation (OPIC), while loans from U.S. commercial banks were guaranteed by the Export-Import Bank. A portion of Freeport Minerals equity investment in Freeport Indonesia was guaranteed by OPIC against loss due to expropriation, war, and currency inconvertibility.

In the typical financial arrangement for a large mine project, the company holding the equity does not directly guarantee the external loans, but negotiates a creditor agreement under which it assumes responsibility for completing the mine by a certain date and providing any additional funds required for completion. Completion is defined in terms of operating at a percentage of capacity for a specified number of months. Failure to meet completion conditions renders the equity holder responsible for making interest and principle payments on the external loans beyond what can be paid by the project. However, once completion conditions are met, the equity holder has no further obligation to the creditors. In most cases, creditors are assured payment of debt from export proceeds from the sale of mine products. This is arranged by having the proceeds paid into a trust fund held by a foreign bank. Otherwise, the creditors would be at the mercy of the host government for payment since most developing countries require that all export proceeds initially be turned over to the central bank. The creditors also require the mining subsidiary to negotiate long-term contracts for the sale of products to consumers. Occasionally, companies holding the equity may agree to purchase a certain amount of the output. In the case of copper, the contracts call for the delivery of a certain amount each year at the market price prevailing at the time of delivery. In some cases, long-term credits may be provided by consumers. In the financing of Freeport Indonesia, $24 million in credits were provided by Japanese firms that had contracted to buy specified amounts of the copper output.

Although the nonrecourse loans described above relieve the foreign equity investor of the obligation to repay loans once completion conditions have been met, there is considerable risk of delay in completion and of technical difficulties that increase the cost of the project well above the original estimate. In the case of Cuajone, actual costs were nearly double the estimate in the original financial plan and these costs had to be covered by the equity investors. Due to serious technical problems requiring the rebuilding of the metallurgical plant, the cost of the Selebi-Phikwe nickel/copper mine was more than double that estimated in the financial plan and, in order to complete the project, the major stockholders (AMAX and AAC) were required to provide about $400 million in subordinated loans (which they are unlikely to recover). In addition, the senior creditors, including a consortium of German banks and the Industrial Development Corporation of South Africa, have suffered substantial losses since even after completion the project did not earn enough to make service payments on the debt, mainly because of low copper and nickel prices. The mine continues to operate with a small operating profit, but it is unlikely to pay any dividends or repay more than a fraction of its indebtedness (Mikesell, 1984b, Chapter 3). Despite low copper prices over the past decade, since 1970 most mines developed by MNCs have not defaulted on their external loans or caused losses for the equity investors, but neither have they been very profitable.

FINANCING STATE MINING ENTERPRISES

Most copper producing SMEs were organized to operate expropriated foreign properties for which the owners were paid only small compensation in relation to their asset value. Therefore, most SMEs came into existence with a relatively large amount of equity and, in some cases, little debt. The few exceptions include Brasiliana de Cobre and some of Minero Peru's mines. However, a number of SMEs have expanded capacity and are continually in need of capital for replacement and modernization. For these purposes SMEs rely mainly on borrowing or reinvested earnings. Governments usually do not provide SMEs with substantial amounts of equity capital beyond that represented by the expropriated assets. With the notable exceptions of CODELCO and CVRD, SMEs have had few profits to reinvest. Hence, the bulk of their capital financing is derived from foreign borrowing in the form of commercial bank loans, equipment credits, and World Bank and Inter-American Development Bank (IADB) loans. SMEs also borrow from domestic credit institutions, mainly government banks.

With certain exceptions, such as equipment credits and loans guaranteed by government export credit agencies, SMEs cannot borrow in the international financial market without a government guarantee. In the past this has meant the rates of interest on loans to SMEs have been lower than the rates on loans to subsidiaries of MNCs, which do not carry a government guarantee. Especially advantageous have been the loans to copper producing SMEs by the World Bank and IADB. The largest loans have been made to Centromin in Peru, CODELCO in Chile, and Gecamines in Zaire. Such loans have represented less than 15 percent of total capital invested in all copper producing SMEs in developing countries. It is unlikely that financing from international development lending institutions will provide a substantial portion of future capital requirements of SMEs. The highest priorities of these institutions are loans for agriculture, various forms of infrastructure, and structural adjustment loans not tied to specific projects. Planned projects by SMEs, mainly in Brazil, Chile and Peru, total about 900,000 mtpy in new capacity, but the realization of these plans will depend on the ability of the governments to obtain external financing and/or to attract foreign investment in the form of joint ventures. Costs of constructing this capacity at an estimated $5,800 per mt would be about $5.2 billion plus another $1.8 billion for smelter capacity, or a total of $7.0 billion (in 1985 dollars). Over a five-year period this would represent an outlay of $1.4 billion per year. Although the amount is small compared to the total capital flow to developing countries in the 1970s, it is considerably larger than annual capital expenditures of SMEs on copper projects over the past decade.

SMEs in most developing countries will be faced with substantial difficulties in obtaining private international financing for large new copper mining projects over the next decade. Most producing countries, including Chile, Mexico, Peru, Zaire and Zambia, have either defaulted on external debt or have renegotiated foreign loans; consequently, their ability to obtain new external private financing is quite limited. However, given the technical and managerial competence of some SMEs, it is conceivable they might be able to borrow without a government guarantee to develop new mine capacity. Foreign lenders might be willing to lend directly to an SME provided service payments were assured by the establishment of a trust fund into which export proceeds would be paid, similar to the arrangements used by subsidiaries of MNCs. However, I know of no commercial bank loans made to copper producing SMEs under these conditions.

The formation of joint ventures between SMEs and foreign mining companies has been regarded as a promising means of financing investment in copper and other metals. Joint ventures have been used successfully in the bauxite-aluminum industry in Brazil and in nickel and coal production in Colombia. However, joint ventures require both the government and the foreign investor to share in financing mine construction. Moreover, foreign investors in joint ventures frequently demand managerial control of the enterprise. The only joint venture in copper between a government and a foreign investor of which I am aware is that between the Overseas Mineral Resources Development Sabah Bdh. (a consortium of Japanese copper importers) and the Government of Malaysia. Minero Peru has been seeking foreign joint venture partners in its copper projects, but to my knowledge it has not been successful.[5]

CONCLUSIONS AND OUTLOOK

The climate for FDI in major Third World copper producing countries ranges from reasonably good for Chile, PNG and Indonesia to poor or unacceptable for Mexico, the Philippines, Peru, Zaire and Zambia. Any FDI in the copper industry in Brazil is likely to take the form of a joint venture with CVRD. Plans exist for FDI in copper projects in some countries not currently significant producers, the most important being in Argentina and Panama (Mikesell, 1983, Chapter 12). However, these investments are not expected to be made in the near future. With an increase in the price of copper to 80 cents per pound, FDI in projects with a total capacity of 350,000 mt are quite likely to be made in Chile over the next decade, especially if a moderate democratic government is established in that country. There are also a number of new projects planned by SMEs in Argentina, Brazil, Chile and Peru totaling nearly 1 million mt, but only those in Chile and Brazil appear likely to be financed in the foreseeable future.

All of this suggests a rather slow growth in Third World copper mining capacity over the next decade, even if prices should rise by 20 percent over 1986 levels. Meanwhile, there are quite a number of planned projects, either for new mines or major expansions of existing ones, in Australia, Canada, Portugal and the U.S. Given the proper price incentives, it is difficult to forecast whether capacity growth will be greater in Third World or in developed countries over the next decade.

APPENDIX 5-1: THE RESOURCE RENT TAX (RRT)

The basic principle of the RRT is to tax the accumulated net present value (NPV) in any year that it is positive. (Accumulated NPV is the present value of cash receipts less all cash outlays over the life of the project.) All taxes, including any accounting profits taxes or royalties paid by the investor, are subtracted in calculating accumulated NPV. The rate of discount used for calculating the accumulated NPV is determined by the government, or may be negotiated with the contractor. It is normally chosen to represent the minimum IRR necessary to attract an investor and includes an allowance for risk above the rate on riskless securities. In the example given in the accompanying table, the discount rate used to calculate the accumulated NPV is 18 percent. A positive accumulated NPV is not realized until the 8th year of operations, at which point an RRT of 60 percent is applied. In the 9th year net cash receipts are all taxed at 60 percent, but in the 10th years accumulated NPV becomes negative because net cash receipts are negative. In the 11th year the accumulated NPV again becomes positive and that amount is taxed at 60 percent, yielding a revenue of 25.

NOTES

1. For a review of the literature see Walde, 1985, pp. 3-33; Mikesell, 1984a, and Mikesell, 1984b, pp. 79-89.
2. Before the Mexicanization program of the 1960s, Anaconda and ASARCO were majority equity holders of Mexico's major copper mines, Minera de Cananea and Industria Minera Mexico, respectively. Currently the Mexican government owns 52 percent of Cananea and 44 percent of Mexicana del Cobre, Mexico's largest copper mines.
3. For an analysis of the technical capacity and policies of SMEs, see Mikesell and Whitney, 1987, Chapter 2.
4. ZCCM has been subjected to strong social and political pressures to avoid labor reductions and to replace expatriates with inadequately trained domestic workers with little experience (Radetzki, 1985, pp. 124-26). Radetzki has provided evidence that governments sometimes use state-owned mineral firms to promote a variety of social goals,

Appendix Table 5.1

The Resource Rent Tax: Hypothetical Example of Tax on Accumulated Present Value

	Net cash receipts	A_t (accumulated present value at 18%)	$T(A_t)$ (tax on returns over 18% threshhold tax rate (T) of 60%)
Expenditures in 1st year of capital outlays	-130	-130	
Expenditures in 2nd year of capital outlays	-100	-253	
Expenditures in 3rd year of capital outlays	-100	-399	
Year of operation:			
1st	100	-371	
2nd	100	-338	
3rd	100	-299	
4th	100	-253	
5th	100	-199	
6th	100	-135	
7th	100	-59	
8th	-50	30	18
9th	100	--	60
10th	100	-50	0
11th	100	41	25
12th	100	--	60

NOTE: In the 2nd year of capital outlays, A_t = -130(1.18) - 100 = -253. In the 3rd year of capital outlays, A_t = -253(1.18) - 100 = -399. In the 1st year of operations, A_t = -399(1.18) + 100 = -371, and so on. In the 11th year of operations, A_t = -50(1.18) + 1-- = 41.

Source: Raymond F. Mikesell (1984c) Petroleum Company Operations and Agreements in the Developing Countries, Johns Hopkins University Press for Resources for the Future, Baltimore, p. 47.

and that such activities tend to increase productive costs (pp. 142-43).

5. Minero Peru and Metallgesellschaft are negotiating a joint venture for a copper mine at La Granja. The Peruvian government is negotiating with a Japanese consortium for the development of another copper mine.

REFERENCES

Crowson, Philip C. (1982) "Investment and Future Mineral Production," Quarterly Review, CIPEC, Paris, April/June.

Evans, Barbara J. (1986) "How to Assess the Staying Power of World Copper Mines," Engineering and Mining Journal, 187:4, April, pp. 32-36.

Mikesell, Raymond F. (1983) Foreign Investment in Mining Projects, Oelgeschlager, Gunn and Hain, Cambridge, Mass.

__________ (1984a) "The Evolving Pattern of Contracts and Corporate-Government Relations in Mining," Research and International Business and Finance, IV: Part B, JAI Press.

__________ (1984b) "The Selebi-Phikwe Nickel/Copper Mine in Botswana: Lessons from a Financial Disaster," Natural Resources Forum, 8:3, pp. 279-289.

__________ (1984c) Petroleum Company Operations and Agreements in the Developing Countries, Johns Hopkins University Press for Resources for the Future, Baltimore.

Mikesell, Raymond F. and John W. Whitney (1987) The World Mining Industry: Investment Strategy and Public Policy, Boston, Allen & Unwin.

Radetzki, Marian (1985) State Mineral Enterprises, Resources for the Future, Washington.

Rosenkranz, R. D., E. H. Boyle, and K. E. Porter (1983) Copper Availability--Market Economy Countries: A Minerals Availability Program Appraisal, BOM Information Circular 8930, U.S. Department of Interior, Washington.

Sassos, Michael P. (1986) "Mining Investment in 1986," Engineering and Mining Journal, January.

Survey of Current Business (1986) "Capital Expenditures by Majority-Owned Foreign Affiliates of U.S. Companies," U.S. Department of Commerce, Washington, March.

__________ (1985) "U.S. Direct Investment Abroad," August.

Development and Outlook

Walde, T. (1985) "Third World Mineral Development and Crisis," Journal of World Trade Law, 19:1, January/February.
World Bank (1983) The Outlook for Primary Commodities, Staff Commodity Working Paper No. 9, Washington.

Chapter 6

GOVERNMENT REGULATIONS AND INTERNATIONAL ISSUES RELATING TO COPPER

Copper mining is subject to more government regulations than most industries, the most important of which have to do with foreign trade and protection of the environment. There are also several important international issues that reflect conflicting national policies of producing and consuming countries. These issues include international action to control world copper prices and loans to SMEs by international development banks. This chapter reviews some recent developments in both government regulations and international policies relating to copper.

INTERNATIONAL TRADE POLICIES

Most industrial countries are heavily dependent on foreign supplies of copper and, therefore, have low or no tariffs or import quotas on concentrates or refined copper. The U.S. is the world's largest copper consumer and prior to 1982 was the world's largest copper producer. U.S. mine copper production declined from 1,538 thousand mt in 1981 to an average of 1,033 thousand mt during the 1983-85 period, and net import reliance as a percentage of consumption rose from 6 percent in 1981 to 27 percent in 1985 (Bureau of Mines, 1986, p. 42). This led to a strong demand by the U.S. mining industry for import quotas or a substantial increase in the tariff on refined copper (currently 1 percent ad valorem). In January 1984 the major U.S. copper producers petitioned the U.S. International Trade Commission (ITC) for import relief in the form of a quota under the Trade Act of 1974.[1] In a July 1984 ITC report, all five commissioners found that increased imports of copper are a substantial cause (or threat) of serious injury to the domestic copper industry, but a majority of the commissioners recommended against quantitative import quotas (U.S. International Trade Commission, 1984, pp. 1-2). This position was upheld by President Reagan. The ITC report cited a number of factors that were more important sources of injury to the U.S. copper industry than increased imports. These included relatively high labor costs; high environmental and energy costs; and the combination of a short-term cyclical decline in the demand for copper and a long-term decline in the rate of growth in demand. The commissioners argued against a quota on the grounds that a substantial increase in the domestic price of refined copper above the world price would result in a flood of imports of semi-fabricated products. According to the commissioners "This process has the potential to destroy 98 percent of the demand for domestically produced copper." (ITC, 1984, pp. 47-52). By increasing the domestic price above the world price, domestic fabricators would be unable to compete with foreign fabricators in either domestic or foreign markets. The demand for domestically-produced metal would, therefore, decline.

The trade policy of the Reagan Administration has been to avoid the use of import restrictions that violate the General Agreement on Tariffs and Trade (GATT) rules or invite retaliation by foreign countries. The Administration has, however, supported retaliatory action against "unfair trade practices" by foreign countries, which include dumping, exporting at prices below full cost, export subsidies, and actions limiting U.S. access to foreign markets. During the 1981-84 period the principal sources of U.S. copper imports were Chile (39 percent), Canada (25 percent), and Peru (8 percent). It would be difficult to show that either Chile or Canada is guilty of such unfair trade practices. In the case of Peru, U.S. imports are produced almost entirely by SPCC, while many Canadian copper mines are also owned by U.S. firms. Legislation has been introduced that would require import restrictions against countries that have large trade surpluses with the U.S., but this is not the case for any of the major copper exporters to this country, except Canada. Since the U.S. and Canada are planning to negotiate a free trade agreement, it is unlikely the U.S. will impose import barriers on Canadian copper as a means of supporting the domestic copper industry. However, there is considerable danger that the Democratically-controlled Congress will legislate import quota restrictions on copper from other countries.

ENVIRONMENTAL REGULATIONS

Copper mining and processing create a number of environmental problems (Mikesell, 1987, Ch. 8). Open pit mines deform the surface of the land and create waste materials containing hazardous substances that pollute water and soil. Water from mining and concentrating operations may contaminate the subsoil and the rivers into which it flows. The most serious problem is the gases produced by smelters, which not only contaminate the air in the region of the smelter but may by creating acid rain affect lakes and vegetation hundreds of miles away. Thus nearly all industrial countries and some developing countries impose environmental regulations on air pollution produced by copper smelters. Pollution abatement regulations and standards differ from country to country and with the location of smelters in relation to populated areas within a country. Although pollution abatement standards are not necessarily more strict in the U.S. than in other countries, such as Japan and the Western European countries, their impact on the copper industry has been the most severe for two reasons. First, a large number of U.S. smelters operating in the 1970s and early 1980s were the old reverberatory type constructed early in the century, while Japanese and Western European smelters were newer and less polluting. Second, U.S. legislation requires the introduction of new technology in the form of electric furnaces, flash smelting processes, or hydrometallurgical processing of ores and concentrates; foreign regulations provide greater flexibility in the methods used to control emissions. By the spring of 1986 only two U.S. smelters not in compliance with EPA standards were in operation, namely, Phelps Dodge's Douglas and Newmont's San Manuel, both in Arizona. The Douglas smelter was shut down in 1987 and the San Manuel smelter is being reconstructed to conform to EPA standards. Since 1981 more than one-third of U.S. smelter capacity has been shut down. Although a few new nonpolluting smelters have been built, low copper prices have limited the willingness and ability of mining companies to invest in new smelters. Eliminating polluting smelters still leaves the problem of pollution from two smelters located in Mexico near the U.S. border. The U.S. has been negotiating with the Mexican government to require installation of sulfur-dioxide controls on Mexican smelters.

Environmental standards in Third World countries are generally lower than those in developed countries. If environmental standards are low or nonexistent, it is likely to be reflected in environmental practices of mining affiliates of MNCs (Gladwin and Welles, 1976). However, in recent agreements with MNCs a few developing countries have adopted strict standards and government supervision of environmental practices in mining. The strictest standards have been established by the PNG government, which are best illustrated by the negotiations for the development of the Ok Tedi gold-copper mine. Despite the fact that the mine is located in a remote area inhabited by Stone Age people in a few isolated villages, the PNG government requested an extensive environmental impact study as a condition for its approval of the project. However, the BHP consortium argued it could not afford large outlays for an environmental study prior to approval of the project. On the basis of limited environmental work done in the feasibility study, the BHP consortium concluded that dumping waste rock into the river beneath the mine and depositing untreated cyanide from the gold-cyanide plant would not cause environmental damage. The PNG government undertook its own studies that refuted these conclusions and final government approval of the project was delayed pending a satisfactory agreement on environmental controls (Pintz, 1984).

The existence of more stringent environmental regulations on mining in the U.S. than are applied in some Third World countries, such as Chile, has been regarded as a source of competitive disadvantage for the U.S. Therefore, a bill was introduced in Congress calling for special import duties that would equalize environmental costs. However, discriminatory import barriers for this purpose would constitute a violation of the GATT and are generally rejected by economists (Walter, 1975). Environmental standards constitute part of the social policies of governments similar to minimum wage, health and safety requirements. Import duties designed to equalize the cost impacts of differing social policies would subject world trade to a maze of discriminatory actions. Moreover, it is exceedingly difficult to make comparisons of environmental costs on an international basis.

Governments are becoming more conscious of the importance of preserving the environment and the gap between environmental standards in developed and developing countries will tend to narrow. The World Bank and other international development institutions mandate environmental impact studies for all projects they finance. The standards set by these institutions are likely to be extended to other mining and metallurgical projects. In addition, subsidiaries of MNCs have introduced pollution abatement technology similar to that used in the industrial countries. This has been done in part because of concern for the welfare of communities in which they operate, and in part because more strict pollution abatement standards are likely to be mandated by the host government in the future. Hopefully, these practices will be copied by the SMEs, which have generally followed more environmentally destructive practices than MNCs.

INTERNATIONAL COMMODITY AGREEMENTS

As discussed in Chapter 4, copper prices have shown a high degree of volatility, more than for any other mineral. This has contributed to erratic changes in investment in new capacity, which in turn has exacerbated price fluctuations. Price fluctuations not related to changes in costs of production contribute to uncertainty and risk for both consumers and producers and often result in misdirected investment and inefficient use of resources. In addition, they create serious balance of payments problems for countries such as Chile, Zaire and Zambia that depend on copper exports for over half their foreign exchange income. Therefore, developing countries have been promoting an international commodity agreement for stabilizing the price of copper as a part of a broader program for stabilizing prices of all primary commodities of major importance to developing countries (Mikesell, 1987, Ch. 6).

The creation of an international mechanism for stabilizing prices of primary commodities has been a principal objective of the UN Conference on Trade and Development (UNCTAD), an agency dominated by representatives of Third World countries. At the

UNCTAD conference in Nairobi in 1976, the Secretariat proposed, with strong endorsement by developing countries, an Integrated Program for Commodities (IPC). The IPC calls for the negotiation of international agreements for ten core commodities of which copper and tin are the nonfuel minerals, plus price stabilization measures for eight other commodities, including bauxite and iron ore. Membership in the commodity agreements would include both producing and consuming countries. In addition, the IPC provided for a common fund of $6 billion, to be subscribed by governments for financing individual agreements. Subscriptions by governments to the common fund have not been made and UNCTAD has had only limited success in promoting the negotiation of additional commodity agreements. At the time the IPC was formulated, international agreements for four of the ten core commodities--cocoa, coffee, sugar and tin--were already in operation. (The U.S. has been a member of the coffee and tin agreements from time to time.) UNCTAD has given special emphasis to negotiating an international commodity agreement for copper. A commodity agreement in copper has also been strongly supported by the Intergovernmental Council of Copper Exporting Countries (CIPEC) whose membership includes the principal Third World copper exporting countries (Chile, Indonesia, PNG, Peru, Yugoslavia, Zaire and Zambia), plus Australia. However, representatives of major industrial countries and Third World producing countries have been unable to reach an agreement on a stabilization program.

Most industrial countries have not enthusiastically supported commodity agreements, but have sought to accommodate the interests of Third World countries in establishing them--provided their objective is true price stabilization and not simply fixing prices at arbitrary levels. True price stabilization means modifying fluctutions above and below the long-run equilibrium price. This may best be achieved by a buffer stock-type arrangement in which the stock would consist of both international money and the commodity. The buffer stock manager would buy the commodity when the world price fell below the estimated long-run equilibrium price by a certain percentage, and sell when the price rose above the estimated price by a certain percentage. The success of a buffer stock operation would be indicated by reestablishment of the initial composition of the stock in terms of money and commodity from time to time over a period of, say, three to five years. The UNCTAD Secretariat and Third World countries favor the use of export quotas to supplement operations of a buffer stock whenever funds are insufficient to maintain a price for the commodity regarded as "fair" or equal to some historical level. Buffer stock operations do not interfere with the freedom of the market, but operate within the context of a free market by influencing demand and supply. Export quotas, on the other hand, restrict normal market operations, and while they might be successful in maintaining a floor price in the short run, they are likely to create large surpluses that will burden the market over the long run.

A major reason for the failure to negotiate a commodity agreement for copper acceptable to industrial countries and Third World exporting countries is the difficulty of formulating a feasible buffer stock program that could operate without the use of quotas. Quite a number of technical studies have been published on the feasibility and cost of a copper buffer stock, but the studies reach different conclusions (Adams and Klein, 1978; Charles River Associates, 1977; Maizels, 1982). Economists have pointed out that it is extremely difficult to project a long-run equilibrium price that could serve as the reference price or target for a stabilization operation. A simulation study by CRA (1977) showed that a buffer stock capable of achieving reasonable price stability is likely to require several billion dollars plus a large copper stock, and that industrial countries are unlikely to contribute resources of the size required for this purpose. A more recent study showed that an effective buffer stock would require about $2 billion, but given depressed prices and uncertainties regarding the future rate of growth in demand, agreement on a reference price around which to stabilize prices would be exceedingly difficult to obtain. For example, Third World producing countries would be unlikely to accept a reference price of 65 cents per pound--the average LME price in 1985. The principal objective of Third World copper exporting countries is to raise the price well above the 1985-86 level, but this cannot be done with a moderate-size buffer stock if the equilibrium price is unlikely to increase much above that level for the next decade. A buffer stock of $2 billion would soon be exhausted if it attempted to raise the price to 95 cents or $1 per pound. Moreover, it is unlikely that the U.S.

and other developed countries would subscribe substantial amounts of money for this purpose. Therefore, it appears that an international copper agreement will not be negotiated in the foreseeable future.

A good example of the problems faced in stabilizing the market price of a mineral by means of a buffer stock operation is provided by the experience of the International Tin Association (ITA). The ITA was successful in supporting the price of tin for a number of years, but did not achieve a significant degree of price stability. In periods of high demand, the price of tin was well above the ceiling price because the buffer stock lacked sufficient tin to prevent the price from rising. During the 1980s the growth in demand for tin was lower than the growth in supply, but the support price was maintained at a level that promoted increased world production (including that by nonmembers of the ITA), and encouraged substitution of other materials for tin in industrial production. The ITA attempted to maintain an unrealistic support price by borrowing hundreds of millions of dollars against tin acquired by the buffer stock, but in October 1985 the stock ran out of funds and the price fell to less than half the support price. Trading in tin on the LME was suspended and a large amount of tin in the hands of the creditors that loaned money to the buffer stock overhangs the market.

An alternative to a buffer stock would be an agreement among copper exporters to establish production and export quotas designed to raise the price to some target level. The OPEC countries were able to maintain high prices for petroleum for nearly a decade; this was accomplished mainly because of the dominant market position of Saudi Arabia and the willingness of a few other producers to cooperate in maintaining export prices by limiting production. These conditions do not exist for any major nonfuel mineral. Experience has shown that producer cartels are unable to maintain discipline among members for very long or even to agree on export quotas.

Despite the economic costs of price volatility and long periods of prices below costs, there appears to be no satisfactory alternative to the operation of competitive forces in the world copper market.

PUBLIC INTERNATIONAL ASSISTANCE TO THE COPPER INDUSTRIES IN DEVELOPING COUNTRIES

Public international assistance to mineral industries of developing countries has taken several forms, including technical assistance from the UN, loans from international lending agencies, and bilateral assistance from governments. Although some assistance has been provided to promote private domestic and foreign investment, the bulk has gone to SMEs. Both developed and developing countries have an interest in promoting the flow of capital and technology to developing countries for expanding mineral output. However, the form and direction this assistance should take is a source of considerable controversy. The U.S. and most other developed countries have favored private enterprise, both domestic and foreign. Private enterprise is widely regarded as more efficient than public enterprise, but this is not always the case. More importantly, private investment is available for Third World mineral investment to a greater degree than investment in infrastructure, agriculture, and social programs. Hence, given the limited amount of funds for public assistance, such assistance can be used more effectively as a catalyst for promoting the flow of private capital to the mineral industries of these countries. This position conflicts with that of the majority of developing countries that favor international assistance to public mining and processing enterprises. The principle of "full permanent sovereignty of every state over its natural resources, including the right of nationalization or transfer of ownership to its nationals" has been an important element in the New International Economic Order (NIEO). The program of action for promoting the NIEO is set forth in a UN resolution stating that efforts should be made to "insure that competent agencies of the UN system meet requests for assistance from developing countries in connection with the operation of nationalized means of production." (United Nations, 1974).

Prior to the mid-1970s the World Bank regarded mineral development as mainly a province of private multinational companies and made few loans to the mining and petroleum sectors. The wave of expropriations of private investment in resource industries in the late 1960s and early 1970s substantially reduced the flow of capital to the mineral industries of Third World countries, and there was concern that growth in producing capacity would not be sufficient to meet growing world demands. An unpublished World Bank staff report (1972) recommended a greatly expanded program of loans by the World Bank Group (World Bank, International Development Association, and International Finance Corporation) for mineral projects. Therefore, the World Bank Group increased its financing of mineral projects in Third World countries, but the total amount for copper projects has not been large.[2] With growing world overcapacity in copper and other nonfuel minerals during the late 1970s and early 1980s, the U.S. became increasingly critical of international public loans to SMEs; U.S. representatives in these institutions have voted against several of the loans that were made. Nevertheless, during the 1980s the World Bank and the IADB made loans for expanding or modernizing the copper capacity of SMEs in Chile, Peru, Zaire and Zambia.

During the Reagan Administration, U.S. opposition to public international loans to SMEs reflected the position of the U.S. mining community and their supporters in Congress that such loans subsidize foreign mineral production, particularly copper, and create world excess capacity to the detriment of the U.S. mining industry. U.S. opposition to public international loans to SMEs raises a serious conflict of principles. On the one hand, the charters of both the World Bank and the IADB state that their loans should supplement the flow of private capital rather than replace it. Since private international capital is available for financing Third World mining activities provided a favorable investment climate exists, it is argued that international development agencies should conserve resources for financing projects for which private capital is not available. However, Third World countries argue that this position violates another provision of the charters that forbid these institutions from discriminating against countries on the basis of their political and economic structure. Thus, if a country's political orientation favors SMEs over foreign private investment for development of mineral resources, it should be respected by the international development banks. The lending institutions have never resolved this apparent contradiction.

It seems unlikely that international development institutions will be inclined to make substantial loans for copper development during the next decade. First, they are mindful of the large overcapacity in the world industry and loans for this purpose would appear to constitute a misallocation of resources. Second, given the shortage of capital available for Third World countries following the debt crisis, other uses of international public funds have a much higher priority. Some U.S. mining executives have criticized balance of payment loans by the World Bank, the IADB, and the IMF to countries producing copper on the grounds that such loans will enable governments to finance an expansion or modernization of SME producing capacity. Although public lending institutions have an obligation to monitor how their loan funds are used to support economic development, it would appear to be highly improper for the U.S. government to attempt to dictate the policies of these institutions in the interest of avoiding a possible adverse impact on a particular U.S. industry.

NOTES

1. The U.S. Trade Act of 1974 provides that companies or labor unions may petition the International Trade Commission for import relief. The ITC then studies the conditions giving rise to the petition and if there is serious injury or threat of serious injury from imports, it may recommend to the President the imposition of tariffs or quotas. The President may reject, accept, or modify the recommendation.

2. Between 1958 and 1983 the World Bank and IDA made seven loans for copper, lead and zinc projects totaling $227 million. In addition, the IFC made loans and investments in the copper, lead and zinc sectors totaling $137 million (Haug, 1983).

REFERENCES

Adams, F. Gerard and Sonja Klein (eds.) (1978) Stabilizing World Commodity Markets, Lexington Books, Lexington, Mass.

Bureau of Mines (1986) Mineral Commodity Summaries 1986, U.S. Department of Interior, Washington.

Charles River Associates (1977) The Feasibility of Copper Price Stabilization Using a Buffer Stock and Supply Restriction from 1953 to 1976, Charles River Associates, Cambridge, November.

Gladwin, Thomas L. and John G. Welles (1976) "Environmental Policy and Multinational Corporate Strategy," in Ingo Walter (ed.), Studies in International Environmental Economics, Wiley, New York.

Haug, Marianne (1983) Unpublished address at the Economics Symposium of the American Institute of Mining, Metallurgical and Petroleum Engineers, Washington, November.

Maizels, Alfred (1982) Selected Issues in the Negotiation of International Commodity Agreements: An Economic Analysis, Geneva, UNCTAD.

Mikesell, Raymond F. (1987) Nonfuel Minerals: Foreign Dependence and National Security, University of Michigan Press for Twentieth Century Fund, Ann Arbor.

Pintz, William S. (1984) Ok Tedi: Evaluation of a Third World Mining Project, Mining Journal Books, London.

United Nations (1974) General Assembly Resolution 3202 (S-VI), New York, May 1.

U.S. International Trade Commission (1984) Unwrought Copper (Report to the President on Investigation No. TA-201-52 under Section 201 of Trade Act of 1974), Washington, July.

Walter, Ingo (1975) International Economics of Pollution, Macmillan, London.

Chapter 7

TECHNOLOGICAL INNOVATION

Without the technological innovations in exploration, mining and processing early in the present century it would not have been possible to supply the enormous demand for copper generated by the modern industrial economy. However, since the 1920s there have been relatively few major technological breakthroughs in the industry, and following the leveling of demand for copper after 1975 there has been little emphasis on R&D by mining companies. Most R&D has been devoted to the improvement of exploration techniques, while new technology in mining has been largely developed by equipment manufacturers and suppliers outside the industry (National Academy of Sciences, 1978, Ch. 1). Over the next two decades, technological advances are not required to expand supply to meet consumer demand, but advances will be needed early in the next century unless other materials are substituted for copper in its major uses. The U.S. copper industry will require the development and adoption of advanced technology if it is to be competitive with other producing countries having higher ore grades and lower labor costs.

The purpose of this chapter is to consider what technological innovations are likely to take place in the industry over the next twenty-five years. To assess the impact of technological innovation on the copper industry over the next quarter century, a study was initiated by the U.S. Bureau of Mines to identify the potential innovations that would have a significant impact on exploration, mining, beneficiation (concentration), smelting and refining over the next two decades. The study also appraised future benefits of these innovations and the degree to which they would be implemented by the mining industry. The study was conducted by a 62-member panel of copper experts from government, industry and academia. Since new technology developed and applied in the U.S. tends to be diffused throughout the world, the findings of this study may be regarded as applicable to the world copper mining industry. In the following sections the findings of the BOM study are summarized for each stage in the production of copper (Weise, et.al, 1983; Sousa, et.al., 1983).

Although an impressive array of technology has been developed, its impact on future copper output will depend upon (1) the expanded use of new technology by mining firms, and (2) further research and experimentation for developing new technology and improving existing technology. The BOM study cited several business environment factors that influence both the diffusion of new technology and R&D for new or existing technology. These include the rate of growth in world demand for copper; governmental policies such as environmental regulations; access to federal lands; macro-economic policies that affect investment incentives; and the organizational structure and behavior of the mining industry. For example, if the copper industry is dominated by firms that choose to devote little capital to grass roots exploration and R&D oriented to exploration, relatively little new exploration technology will be used or developed. The panel of experts based their optimistic and pessimistic cases in large measure on the degree to which favorable business conditions will prevail over the next twenty years.

EXPLORATION

The best examples of high-tech applications in the mining industry are found in the exploration stage. Deposits that can be located by surface observation and sampling have virtually disappeared. Exploration teams include not only geologists, but chemists, physicists, and other science specialists that employ indirect methods for locating deposits based on geologic models, geochemistry, geophysics, telegeologic or remote sensing techniques, side-looking radar imagery or photography, and multi-spectral satellite scanning (LANDSAT). The data sets gathered by using these technologies are integrated and analyzed by sophisticated computer models. Geochemical exploration measures chemical properties of rock, soil or sediment in streams, lakes, glacial debris, and airborne volatile materials. Exploration geophysics uses electronic equipment that can detect subtle contrasts in such physical properties as specific gravity, electrical conductivity, heat conductivity, seismic velocity and magnetic susceptibility as aids in locating deposits in the earth's subsurface. Remote sensing techniques that measure geologic properties from aircraft provide insights into complex structures in the bedrock. Side-looking radar imagery is used where conventional photographic methods are inadequate. LANDSAT imagery provides information on potential mineral deposits as well as ground water conditions. Materials collected in exploration are subjected to sophisticated laboratory analysis that includes chemical analysis, color-comparison, X-ray microanalysis, electron microprobe, and neutron-activation analysis. The information from these various sources is integrated and analyzed by computerized data processing and display techniques.

There are four stages in the exploration sequence beginning with the appraisal of large regions to select areas with favorable mineralization (stage 1). This appraisal is followed by the reconnaissance of favorable regions in the search for target areas (stage 2). Once a target area has been located, detailed surface investigation is made (stage 3) and, if warranted, three-dimensional physical sampling is conducted (stage 4). A discovery cannot be declared mineable until there is a detailed economic analysis of potential costs and revenue from the exploitation of the deposit. Such analysis must be based on physical data derived from exploration. Each of these stages requires the application of a number of techniques, especially since exploration for copper deposits is increasingly aimed toward potential areas where deposits are well below the surface. The BOM study identified and ranked in order of their likely impact on copper discoveries the following nine technologies:

*1. Integration of Multi-Disciplinary Data Sets (geochemical, geophysics, geologic) through improved data processing and display techniques.

*2. Geologic Models, i. e., better understanding of orebody genesis.

3. Improved Use of Airborne Systems, i.e., deeper penetration.

4. Improved and Cheaper Drilling Techniques.

5. Field Use of Computers and Computer Methods, i.e., on-site analysis.
6. Improved Electrical Geophysical Methods, i.e., deeper penetration.
*7. Hydrogeochemical Methods for Locating Deep-Lying Deposits.
8. Computer-Based Analysis of Satellite and Airborne Surveys.
*9. Multi-Spectral Satellite Scanning Data (LANDSAT).

The above technologies were identified and assessed by a panel of experts. Those denoted with an asterisk were described and evaluated in the BOM study and are summarized in the following paragraphs.

Integration of Multi-Disciplinary Data Sets (IMD)

The panel gave the highest ranking to data processing and display technology rather than to technology for gathering exploration data. IMD applies mainly to stage 3 of the exploration sequence at which point the decision is made whether to conduct detailed three-dimensional sampling--a process requiring large expenditures. Appraisal depends heavily on the integration and interpretation of large quantities of data relating to the geology, geochemistry, and geophysics of the target deposits. Computer technology and graphics make it possible to plot multi-element data and integrate geochemical with geological and geophysical data. This requires the development of advanced statistical analysis and computer modeling. Although considerable progress has been made, the panel of experts identified three areas where additional R&D are needed for improving IMD technology: (a) the development of hardware and software for processing large volumes of deposit data; (b) the development of means to provide graphic imagery in thirty to sixty seconds; and (c) improved graphic display methods permitting three-dimensional representation. Under the most optimistic case, the panel projected a fairly rapid growth in the use of IMD technology and that after 1990 it would be used in the discovery of up to 800,000 mtpy of copper. Under the most pessimistic case, little use would be made of this technology.

Geologic Models

Geologic models are important during both the reconnaissance and target investigation stages of exploration. Different types of mineralization and geologic environments call for different geologic models that portray the expected structure of the deposits. Although considerable progress has been made in developing and using geologic models in exploration over the past twenty-five years, several problems requiring future research were cited by the panel. These include (a) the refinement of models in terms of a deposit's relationship to other rocks in the surrounding country; and (b) a better understanding of tectonic and thermodynamic processes in the creation of deposits. Plate tectonics is playing an increasing role in exploration. For example, several South American deposits have been discovered on the basis of information found on the west coast of Africa to which eastern South America is believed to have been attached.

Hydrogeochemical Methods

Five of the nine technologies designated by the panel are for collecting data on the nature of deposits. These include improvement of airborne systems for detecting mineralization below the surface; improved drilling techniques; improved electrical geophysical methods for achieving deeper penetration; and hydrogeochemical methods for locating deep-lying deposits. The latter technology requires special explanation. Water penetrates below the earth's surface, sometimes to a depth of two kilometers, and dissolves minerals with which it comes into contact. When certain elements are concentrated in the water above their natural concentration, an analysis provides a means of detecting the presence of copper deposits well below the surface. However, there is a need to develop better methods of water sampling, of measuring, and interpreting the combinations of elements found in the water to reveal the character and location of deposits. The panel of experts believed these problems would not be solved until the latter part of the 1990s at the earliest. Furthermore, hydrogeochemical methods, which rank 7th in order of potential impact, are expected to contribute to the discovery of significant quantities of copper.

Multi-Spectral Satellite Scanning (LANDSAT)

In copper exploration, LANDSAT may be used for rapid reconnaissance, especially in remote areas with limited access for traditional remote sensing devices. Interpretation of LANDSAT images is based on the recognition of tonal contrasts, shapes, and patterns, using standard geological techniques adapted to wave bands. The experts identified several problems that need to be solved before LANDSAT can become fully useful as an exploration technique, the most important of which is the improvement of the image resolution. Other technical problems cited were (a) development of better means of obtaining information about rock composition and below-surface deposits; (b) a method to overcome the hindering effects of weather and vegetation; and (c) less costly data acquisition equipment. Some of these problems may not be solved until after the turn of the century.

The experts believe that LANDSAT will play a smaller role in future copper discoveries than any of the other nine technologies listed above, and that satellite scanning will be restricted to the discovery of certain types of surface as opposed to deep-seated deposits. The pessimism of the panel of experts regarding the contribution of LANDSAT is somewhat surprising in view of the fact that it has already been successfully employed in minerals exploration and that continual advances in the technology are taking place.

MINING TECHNOLOGIES

Although a number of technological innovations have been made in copper mining during the past half century, there have been no major breakthroughs rivaling the development of open-pit mining of low grade ores that occurred at the turn of the century. Quite a number of advances have been made in underground mining, including improved drilling equipment, mechanized ore loading and transporting, improved blasting agents, and several improvements in equipment for shaft sinking, hoisting, mine ventilation, and safety. However, over 80 percent of U.S. copper mine production and about two-thirds of world production is open pit. Hence, the greatest contribution to mining productivity will come from technology oriented to surface mining. In the U.S. in particular, the competitive position of the industry will depend heavily on the economical mining of increasingly lower grade deposits.

In view of these factors, the panel identified and ranked in order of estimated impact on U.S. copper production, the following technologies:

1. Portable In-Pit Crushing/Overland Belt Conveyor Systems (continuous mining; elimination of trucks).
2. Improved Drilling Techniques (Laser drilling).
3. On-Line (automated) Production Monitoring and Control (dispatching).
4. In-Situ Solution Mining.
5. Improved Dump Leaching (oxygen/air injection).
6. Nuclear Explosive Mining.
7. Borehole Mining (slurry mining).

Overland Belt Conveyors (OBC)

The highest ranking for the seven technologies selected is given to ore crushing in the pit and overland belt conveyors (OBC) that transport mined material to processing plants, thus eliminating hauling by trucks. This technology becomes especially important for transporting lower grade ores extracted from deep mines, with a consequent reduction in labor, fuel and equipment costs. OBC has been successfully used in two major open-pit copper mines in the U.S. and many mines could be converted to this form of continuous mining. The experts cited three problems that need to be overcome before OBCs will be more widely employed in mines: (a) the systems need to be made more flexible so they can be readily moved within the pit; (b) they need to be designed to minimize disruption of operations in the event of breakdown; and (c) they need to be more reliable and to haul larger loads. The experts regard the first problem as the most difficult to overcome and it will probably not be solved until the end of the century at best. Again, this assessment may be too pessimistic since the technology is currently being introduced in U.S. copper mines.

On-Line (Automated) Production Monitoring and Control (Dispatching) (OPMC)

OPMC is a computerized mine monitoring and control system designed to minimize shovel time;

minimize truck queing time; and achieve optimal route selection. OPMC has been implemented in a number of mines so the basic technology exists, but is in need of improvement. Nevertheless, the panel of experts did not foresee the adoption of OMPC by mines producing as much as 1.6 million mtpy until after the turn of the century. Future use of OMPC, which applies mainly to the integration of shovel with truck movements, may be limited to the introduction of in-pit crushing and overland belt conveyors.

In-Situ Solution Mining (ISM)

ISM involves the following basic steps: drilling of wells into the orebody; injecting a liquid (or a mixture of liquid and gas) to dissolve copper out of the host rock; recovery of the pregnant solution; purification and recovery of metals from the pregnant solution; and recycling of the barren solvent.

With ISM no movement of solids occurs since only the liquid moves through the rock and the metal values are leached from the ore by solutions pumped into the deposit. The metal-bearing solution is then treated on the surface by hydrometallurgical techniques to recover copper. ISM could significantly reduce capital and operating costs of producing low-grade ores and provides the possibility of equalizing domestic costs with those of foreign producers using conventional technology to mine high-grade ores. ISM also avoids many of the environmental problems associated with conventional mining and metallurgical operations.

Although there has been limited use of ISM by U.S. copper companies with varying success, widespread commercial implementation will require the solution of several problems cited by the experts. These include (a) improved methods of creating and assessing deposit permeability; (b) better control of solution flow through the orebody; (c) greater metal loading in the solution; (d) the avoidance of metal loss through dilution and precipitation; and (e) improved multi-metal recovery. There also needs to be developed suitable environmental controls in handling the solution. The experts estimated a 50 percent probability that most problems will be solved by the end of the century and that after the year 2000 in situ mining will make a significant contribution to U.S. copper supplies ranging from 100,000 to 400,000 mtpy. One of the experts advanced the view that in situ mining will become the ultimate mining method since as ore grades decline, mechanical methods of handling bulk material will be too costly and require too much energy to be economical.

BENEFICIATION (CONCENTRATION)

The principal beneficiation process used in the U.S. copper industry consists of first crushing and grinding the ore to a fine size and pumping the particles in the form of slurry into flotation cells where the copper-bearing particles are attached to air bubbles rising through the slurry to produce a "froth" which contains the copper concentrate. The remaining particles settle to the bottom of the flotation cells and are discarded as "tailings." This traditional process was developed in an era when ores were richer. The declining ore grades in the U.S. (which may average only 0.3 percent by the year 2000 as contrasted with 0.9 and 0.5 percent in 1950 and 1980 respectively) will require technological advances in beneficiation if the U.S. industry is to remain competitive with foreign producers. The experts identified and ranked the following technologies in order of the estimated impact on copper production:

1. Autogenous and Semi-Autogenous Grinding.
2. Computer-Controlled Mills/Flotation and Grinding.
3. Flotation of Oxide Minerals in Copper.
4. Improved Dump and Heap Leaching Techniques.
5. Selectivity of Beneficiation Process.

Autogenous and Semi-Autogenous Grinding

This method of grinding reduces (semi-autogenous) or eliminates (autogenous) the use of grinding media, such as steel balls or rods, so that the ore is essentially self-grinding at the mills. The technology economizes on the use of power and grinding material and results in lower capital and operating costs. The experts identified several problems that must be solved before autogenous/semi-autogenous grinding methods can become more widely used. These include (a) more durable wall liners and mechnical/electrical components in the grinding mills, and (b) new chemical reagents that promote ore degradation. However, the experts were not overly optimistic that the critical problems pertaining to this technology would be solved during the forecast period. At the current level of technology, the panel forecast a usage rate by 2005 ranging from 300,000 mtpy (pessimistic) to 900,000 mtpy (optimistic) at the 75 percent level of certainty. A somewhat larger usage is projected by 2005 if the probems identified above are solved. New plants are better candidates for this technology than by modernizing and retrofitting existing facilities.

Automatic Process Control of Concentrators

This technology involves a comprehensive computer-controlled system for both the grinding and the flotation processes. Cost savings are realized in crushing and grinding operations by greater throughput and metal recovery, while flotation control systems improve metal recovery and reduce reagent consumption. Substantial savings in operating costs are reported by plants that have implemented this technology. The panel identified several problems that need to be solved for the technology to reach widespread commercial use. These include (a) reducing the difficulties and costs of retrofitting existing plants; (b) improving instrumentation reliability; and (c) adjusting the system to changes in grinding characteristics. The panel estimated there was a 75 percent probability that at least 700,000 mtpy of recoverable copper would be processed through automated mills by 2005, assuming an optimistic business climate and no further major technical breakthroughs. This would probably amount to between one third and one half of U.S. capacity. The usage was forecast to increase to a maximum of 1.1 million mtpy if the major technical problems are solved by 2005.

Improved Selectivity of the Beneficiation Process

Technology in the form of improved specific collector reagents and grinding would lead to better separation of copper sulfide minerals from other sulfides and recovery of byproduct mineral values. Several potential advances were identified that would improve the selectivity of the beneficiation process, but the panel felt significant changes would not occur for at least 20 years. At the 75 percent level of certainty (under an optimistic scenario), the panel forecast these improvements would be introduced during the 1990s and by 2005 would be adopted by up to half of U.S. mills.

SMELTING TECHNOLOGY

Smelting is the process of separating copper metal from the impurities in which it is chemically combined or physically mixed. There are two basic methods of extracting copper from concentrates--pyrometallurgical processes (to which the term "smelting" has traditionally been applied) and hydro-metallurgical processes. Until recently, the most commonly used procedure in the U.S was smelting copper concentrates with suitable fluxes in a reverberatory furnace following a preliminary roast to eliminate impurities such as arsenic and sulfur. In the reverberatory furnace the lighter impurities combine and float to the top as slag to be skimmed off, while the copper, iron, most of the sulfur, and any contained precious metals form a product known as matte, which is drawn from the lower end of the furnace. The molten matte is transferred to a converter in which air flowing through the matte burns off the sulfur and copper. This method has the disadvantage of high energy consumption and of producing a large amount of sulfur dioxide (SO_2) and other harmful gases. EPA regulations have largely ended its use in the U.S. Other pyrometallurgical processes, such as flash smelting, avoid some of these disadvantages.

Hydrometallurgical processes are based on producing a liquid containing copper from leaching and extracting copper from the solution by means of chemical or electro-chemical reactions.

The panel of experts selected eleven smelting technologies ranked by their expected relative importance to future U.S. copper production:

1. Flash Smelting.
2. Continuous Smelting (Noranda/Mitsubishi/ Outokumpu).
3. Fugitive Control.
4. Solvent Extraction (electrowinning).
5. CLEAR Process (hydrometallurgy).
6. Refactories (increased on-stream time by better maintenance and/or improvements in corrosion-erosion characteristics).
7. CYMET Process (hydrometallurgy).

8. Improved Precious Metals Recovery.
9. Congeneration of Power from Smelting.
10. Advanced Byproduct Recovery System.
11. AMAX Blast Furnace Smelting.

Flash Smelting Process

The flash smelting process was first developed in Finland in the early 1950s and is widely used around the world. The Finnish technology is called the Outokumpu process, but several newer flash furnace processes, including Noranda and Mitsubish, have been developed over the past thirty years. Three flash smelters have been constructed in the U.S.--ASARCO's Hayden, Phelps Dodge's Hidalgo, and Kennecott's Chino. About two-thirds of the new smelters constructed in market economies since 1970 use this process and it appears likely it will become the most widely used process in the U.S. In this process concentrates and fluxes are injected with preheated oxygen-enriched air into a furnace and smelting temperatures are attained as a result of the heat released by the flash combustion of iron and sulfur. Products include matte, slag and offgases similar to those produced in the reverberatory furnace, but are of smaller volume and contain a higher concentration of sulfur dioxide. The advantages include greater energy efficiency; the capture of SO_2 in the form of sulfuric acid; and adaptability to computerization and automated process controls.

Continuous Smelting and Converting Process

In this technology the smelting and converting processes are combined in a single furnace or in a functionally connected series of furnaces. Two such processes have been developed, the most important being the Noranda continuous smelting technique. Kennecott's Garfield (Utah) smelter built in 1974 is the only plant in the U.S. using this technology. The advantages include reduced operating and capital costs by combining several processes in a single vessel, and sufficient concentration of exhaust gases for economic production of sulfuric acid. Ideally, the process should convert copper concentrate into blister, but thus far it has been necessary to upgrade the blister copper by further treatment in a converter.

Several major problems were cited by the experts that limit the adoption of this technology. They include (a) eliminating the need for converters; (b) reducing gas volume; and (c) lowering the level of energy consumption. The panel was not optimistic regarding the solution of all these problems by the end of the century, but estimated the process would be adopted by three or four domestic smelters over the next two decades.

Hydrometallurgical Processes

In the pure hydrometallurgical technology, concentrates are leached in an acid solution and the copper separated by chemical means (or by electrolysis) to produce blister or refined copper. Three pure hydrometallurgical processes for producing sulfide ores have been developed in the U.S.--the Cymet process developed by Cyprus Mines; the Arbiter process developed by Anaconda; and the CLEAR process developed by Duval Corporation. Only limited commercial production has taken place using these technologies. The processes have the advantage of eliminating smelting and converting; avoiding emission of sulfur dioxide into the atmosphere; producing elemental sulfur instead of acid; and adaptability to small-scale processing plants. However, extended commercial use of these pure processes would require technologies to reduce the relatively large energy consumption; to sufficiently remove impurities to produce cathode quality copper; and to recover byproducts. The panel was not optimistic regarding the solution of major problems and/or adoption of pure hydro-metallurgical processes on a broad scale by the year 2005. Such processes may become quite important in the longer run.

In other hydrometallurgical processes, the ore is first concentrated and roasted to produce an oxidized iron/copper sinter which is then leached. The most common method of recovering copper from leach solution is by precipitation with scrap iron. This process produces copper precipitates (commonly known as cement copper), which is smelted and processed by conventional means. Another method involves passing the copper-bearing fluid through a series of solvent extraction sets using ion-exchange reagents to concentrate the copper in the solvent, which is then treated electrolytically in electro-winning cells. A fairly high-grade copper cathode may be produced by this method. Leaching methods are widely used in the African copper belt and in Chile. Currently about 20 percent of U.S. copper is produced by leaching, but only a portion involves the use of solvent extraction-ion exchange technology.

The panel of experts cited several problems that need to be solved before the solvent extraction-ion exchange technology becomes widely used. These include (a) the development of extractants that have a higher copper loading capacity; (b) the design of more efficient mixers and settlers; and (c) the reduction of lead levels in the cathodes. The experts believe that at the 75 percent confidence level and assuming technical difficulties are overcome, this technology will account for as much as 300,000 mtpy in the U.S. by 2005.

REFINING

Copper produced in nearly all smelting processes and in some hydrometallurgical processes contains impurities and byproducts. Electro-refining consists of smelting copper in a furnace to remove the principal impurities, and casting the copper into anodes for electrolytic refining. The anodes and thin copper starting sheets, or cathodes, are suspended in tanks containing a solution of copper sulfate and sulfuric acid. An electric current is passed through the solution dissolving copper from the anodes and depositing it in refined form on the cathodes. Periodically the cathodes are removed from the tanks and the pure copper removed. The sludge collecting on the bottom of the tank contains valuable byproducts which are recovered separately.

The panel examined the following refining technologies ranked in order of their estimated impact on the U.S. copper industry:

1. Automation of Tank House Operations.
2. Periodic Current Reversal.
3. Fluid Bed Electrode.
4. Electrode-Slurry Process.

Automation of Electro-refining

Electro-refining of copper is labor intensive requiring about 2.75 man hours per mt of metal produced. Therefore, a number of automated systems have been developed. One obstacle to adoption of automated systems is that refining constitutes less than 10 percent of total production costs so that potential savings may be low in relation to costs of equipment and instruments. At a 75 percent probability level, usage by 2005 is expected to total no more than 250,000 mt. The experts believed that future investment by copper companies will be concentrated in other phases of production where payback might be larger and faster.

The panel did not foresee a major role for the other technologies listed, either because of time and difficulty required to solve the problems, or because of limited economic advantages compared to those from the introduction of other technologies. Perhaps the most interesting of these technologies is the electro-slurry process consisting of grinding ore into fine particles which are fed into an electro-slurry cell. The cell combines the two steps of leaching and electrowinning by direct extraction of copper on the cathode. This technology would eliminate the need for smelters and would facilitate extraction of copper from complex ores. The concept has been partially tested in the laboratory, but the panel was pessimistic about the likelihood of solving the most critical problems, which include (a) adoption of ultra-fine grinding technology and (b) improved byproduct recovery. Substantial R&D outlays would be necessary to make the process viable, but the panel believes such expenditures would yield greater returns when applied to other stages of copper production.

BENEFIT-COST RATIOS FOR COPPER TECHNOLOGY

In order to appraise the potential economic advantages of the various technologies assessed by the panel of experts, benefit-cost analysis was applied to eleven for which sufficient data were available. A number of important technologies had to be excluded from analysis, including exploration technologies, overland belt conveyors, automated process control of concentrators, and improved selectivity of the beneficiation process. The analysis was applied to the period 1985-2005, but changes in economic and physical conditions beyond 2005 might produce quite different results. For example, in the case of in-situ solution mining (the only mining technology to which benefit-cost analysis was applied), the ratios for different scenarios and magnitudes of operations were all less than one. Later in the next century when and if higher copper prices permit mining of much lower and deeper ore grades, in situ solution mining might have advantages over conventional technologies. For the beneficiation technologies, semi-autogenous grinding was found to have benefit-cost ratios ranging from 9 to 11. Among the smelting technologies, continuous smelting and converting technology showed benefit-cost ratios ranging from 1.7 to 1.9, while some hydrometallurgical processes had ratios only slightly higher than 1. Finally, in refining technology, the automation of the electro-refining process was found to have benefit-cost ratios of 4 to 16.

The BOM study did not provide any estimate of the potential economic benefits of developing and implementing all of the most promising of the technologies examined, or make an overall appraisal of how these technologies might affect the competitive position of the United States in the world copper industry, or the competitive position of the world copper industry in relation to other materials.

RECENT APPLICATIONS OF COST-REDUCING TECHNOLOGY

There are few cases where a large national industry has suffered more severe impacts over a two-year period than the U.S. copper industry. Between 1981 and 1983 output declined by almost one-third and nearly all copper mining firms registered losses on their operations. However, the response among firms in the U.S. differed significantly. Some, like Anaconda, went out of the copper business entirely. Virtually all U.S. copper producing companies reduced output and closed high-cost mines, some permanently. Most reduced capital expenditures, but Kennecott-Sohio and Phelps Dodge have increased their expenditures for modernization and development of new facilities. Phelps Dodge is unique among U.S. mining companies in having increased its copper output during the 1980s. In 1985 its output was 373,000 mt, 20 percent above its previous record of 312,000 mt set in 1979. Phelps Dodge overtook Kennecott to become the largest U.S. copper producer. This was accomplished while one of Phelps Dodge's major mines, the Ajo Arizona mine, was shut down. However, Phelps Dodge's smelter capacity was substantially lower than the 1981 level due to the closure of its Morenci smelter. Phelps Dodge costs were one-third lower in 1985 than in 1981 (without adjusting for inflation), and the company earned a profit on its operations after three consecutive years of losses. Although Phelps Dodge's wage and employment reductions and management improvements were in part responsible for the cost savings, considerable credit must be given to the introduction of new technologies. These included installation of computerized truck dispatching; extraction of pure copper from liquid materials without smelting by means of solvent-extraction electrowinning; placing the Tyrone mine on twenty-four hour-per-day production; and building a new flash smelter at Hidalgo.

Kennecott-Sohio announced in December 1985 that it would undertake a $400 million modernization of its Utah Copper Division which will have capacity of 170,000 mtpy, plus gold, silver and molybdenum byproducts. The modernization includes installation of in-pit ore crushers and a conveyor belt system to eliminate all rail hauling of ore to processing plants. New state-of-the-art grinding facilities will be located near the mines to replace inefficient grinding mills at existing plants ten miles from the mine. An ore slurry pipeline will be constructed to provide low-cost transport of ore to flotation facilities at existing plants. According to a Kennecott-Sohio official, the modernization coupled with labor cost reductions, will make the Utah Copper Division a low-cost domestic producer competitive in international markets at current prices.[1]

At its Chino operation in New Mexico, Kennecott formed a partnership with Mitsubishi of Japan in a renovation project that involved installation of a new concentrator and building a flash smelter. Overall operating costs were reduced by about 31 cents per pound. A continuously operating furnace (using a patented Kennecott process) replaced multiple converters operating intermittently. This process also uses computer control and permits reduction of gas emissions. (Kennecott's two-third interest in the Chino mine complex was sold to Phelps Dodge in September 1986.)

During the period 1982-1984 U.S. mining industry spokesmen were unanimous in arguing that foreign competition would destroy the domestic industry unless protected by high import barriers. Recently, some U.S. mining firms have decided to meet foreign competition through modernization and reduction of costs.

NOTES

1. The Utah Copper Division was suspended in early 1985 due to large operating losses, but will be restored to production following modernization (World Mining Equipment, 1986, p. 6).

REFERENCES

National Academy of Science, (1978) Technological Innovation and Forces for Change in the Minerals Industry, Washington.

Sousa, Lewis, Alfred Weise, and Nicholas J. Themelis (1983) "Assessment of Projected Technological Innovation in the U.S. Copper Industry," Mining Engineering, 35:9 and 10, September 1983, pp. 1271-1273 and October 1983, pp. 1401-1405.

Weise, Alfred, Nicholas J. Themelis, and Nellie E. Guernsey (1983) Technological Innovation in the Copper Industry, Bureau of Mines, U.S. Department of Interior, Washington, March.

World Mining Equipment (1986) "Modernization at Utah," 10:2, February.

Chapter 8

THE FUTURE OF THE WORLD COPPER INDUSTRY: SUMMARY AND CONCLUSIONS

The world copper industry has been regarded as "depressed" for the past decade. Some U.S. mining officials believe the national industry will never again be profitable unless the government restricts imports from the developing world and international development agencies cease making loans to Third World copper producing countries. Spokesmen for Third World copper exporters not only regard such actions as discrimination against poor countries, but believe the world industry cannot recover without international action to raise prices. Neither of these recommended actions seems likely to occur nor provides a viable solution to the problems of the industry. The world copper industry will again become profitable, but the exact nature and timing of the adjustment process is difficult to forecast.

The current problems in the industry have arisen mainly from the creation of substantial overcapacity in the 1970s and the secular decline in the growth rate of world demand. The effects of these factors on prices were exacerbated by the 1982-1984 recession. Special problems in the U.S. industry resulted from the sharp rise in the value of the dollar in terms of foreign currencies, and the necessity of either making large capital outlays to meet EPA pollution abatement standards or shutting down smelter capacity. Special problems for at least some Third World countries have arisen from their inability to obtain foreign exchange for plant maintenance and modernization of SMEs. This occurred following the Third World debt crisis in the early 1980s. Also, most Third World SMEs have been unable to offset lower real copper prices by increasing productivity. However, some of them have regained profitability despite low copper prices by modernization and large devaluations.

LONG-RUN DEMAND PROJECTIONS

Projections of the rate of growth in world demand for copper between now and the end of the century range from 1 percent per year to more than 2.5 percent. Statistical projections derived from regression analysis based on historical time series data and forecasts of economic variables, such as industrial production and GNP, are inadequate indicators of future demand because they do not take into account increases in the pace of conservation and substitution of materials competitive with copper. Although relative prices have some effect on substitution and conservation, both are heavily affected by technological developments. In addition, the composition of GNP has shifted in favor of service and high value-added goods that economize on metals.

An important element in forecasting demand to the end of this century is estimating the growth in the industrial sectors that are major users of copper. This has been done for the U.S. by the BOM, but not for other countries. Over half of U.S. copper consumption in 1983 was for producing electrical and electronic products. The BOM forecasts copper demand for these products for the year 2000 to range from 1.4 million mt to 2.5 million mt; this translates into annual rates of growth of 2.9 to 4.5 percent, respectively. The differences between the low and high forecasts of demand arise from alternative projections of rates of growth in GNP and industrial output, estimates of shifts in industrial composition, and estimates of substitution of other materials. Judgmental factors play a significant role in projecting demand and explain the wide range in annual growth rate forecasts among investigators presented in Chapter 3. Rates of growth in <u>consumption</u> in Western Europe and the U.S. over the past fifteen years have been similar, but the rate of growth in consumption for Japan over the period has been nearly three times that for other industrial countries; the growth rate in consumption for developing countries has been nearly five times. While developing countries consumed only 21 percent of total market economy copper output in 1984, their proportion of total consumption will rise very rapidly if they should continue to grow at an annual rate of 5.6 percent (as contrasted with 1.2 percent for industrial countries). However, real GNP growth in developing countries declined sharply between 1980 and 1984 and there is reason to believe lower rates will continue over the next ten years. This will reduce the rate of growth in copper consumption in the developing countries. In addition, consumption in these countries will be reduced by the same technological developments that have reduced consumption in the developed countries.

There are some developments tending to increase the use of copper or moderate the slowdown in consumption growth that occurred after 1970. One is the expected expansion of data communications and cable television. Another is the partial restoration of the position of copper in the building wire market in competition with aluminum (which is a potential fire hazard). Still another is the penetration of copper into the market for tubing in competition with steel and aluminum (Bureau of Mines, 1985, pp. 20-24). On the other hand, substitution of materials, such as aluminum, plastics, ceramics and optic fibers, in major uses for copper appears likely to accelerate in the future, thereby overwhelming the expansion factors (Bureau of Mines, 1985, pp. 21-23). For these reasons it seems unlikely that the rate of growth in consumption during the remainder of the century will exceed that for the 1970-1984 period (about 2 percent per year). Given the range of judgmental factors in projecting copper demand and the range of forecasts based on econometric models, it is not possible to provide a precise estimate of the rate of growth in consumption for the market economy countries. For purposes of projecting future capacity and prices, I shall assume an annual growth rate of 1.5 percent from 1985 to 2000.

COMPETITIVE STRUCTURE OF THE WORLD MARKET

Assuming no increase in U.S. import barriers on refined copper, the world market will be increasingly competitive and U.S. producer prices will tend to follow those on the commodity exchanges more closely as U.S. import dependence continues to grow. Over the longer run, U.S. and Canadian output may increase moderately, but the ratio of their output to world output will decline. Western European mine copper output is likely to decline absolutely, while Australian output is likely to increase.

Production of mine copper in the U.S. is likely to be concentrated in no more than ten firms by 1995, with over half of it produced by Phelps Dodge and Kennecott. Smelter capacity is likely to be concentrated in five smelters, namely, ASARCO's Hayden (Arizona); Phelps Dodge's Chino (New Mexico) and Hidalgo (New Mexico); Inspiration's Miami (Arizona); and Magma's San Manuel (Arizona). The first three are flash-type smelters while the fourth is an electric furnace. Magma's smelter is scheduled to be rebuilt to comply with federal and state air quality standards. Total smelter capacity of the five is about 720,000 mt. Although additional flash smelters may be built, there is a trend toward the use of leach-electrowinning technology. For example, Magma Copper (a subsidiary of Newmont Mining) is building a solvent extraction-electrowinning plant at San Manuel and Phelps Dodge is constructing such a facility for its Morenci, Arizona mine. Both Magma and ASARCO are considering in situ leaching as a means of producing low-cost copper from their properties in Arizona.[1] Some U.S. concentrates will continue to be exported for a time until new smelter facilities are built.

THIRD WORLD PRODUCTION

In 1984, 59 percent of mine copper output in the market economies and 54 percent of mine capacity was in the developing countries. These percentages are likely to rise during the remainder of this century and beyond, but how rapidly depends in large measure on the volume of FDI and the ability of SMEs to obtain financing for capacity expansion and modernization. Capacity expansion by foreign investors planned over the next five years totals 530,000 mt, while planned capacity expansion by SMEs over the same period totals about 900,000 mt. However, these plans are contingent on an increase in copper prices and the availability of loan financing. Forecasts of actual capacity increases in Third World countries differ substantially among investigators. The largest expansion is projected by CIPEC (see Table 2.5)--an increase of 900,000 mt in Third World capacity between the end of 1984 and the end of 1990. The Phelps Dodge (1985) research staff projects an increase in Third World capacity of 500,000 mt between mid-1985 and the end of 1989. The World Bank staff study (Takeuchi, et.al., 1986, p. 126) projects increases in Third World capacity of 625,000 mt between the beginning of 1985 and the beginning of 1990, and of 800,000 mt for the eleven-year period 1985-1996. In each of these projections, Chile's CODELCO accounts for two-thirds or more of the increases. In the Phelps Dodge and World Bank projections, relatively little increase in Third World capacity through 1990 is based on foreign investment, while in the World Bank projection for 1996 foreign private investment in Chile accounts for the bulk of the increase after 1990.

In the World Bank projection there is a decline of 90,000 mt in Zambia's capacity between 1990 and 1996, offset in part by a modest growth in Philippine and PNG capacities. The projected 90,000 mt rise in Mexico's capacity between 1985 and 1990 reflects a program well underway for that country's mixed government/private domestic mining companies. An analysis of the capacity projections for individual countries made by the World Bank suggests that increases by SMEs between 1985 and 1996 may not be larger than those based on foreign private investment. Currently the percentage of Third World copper output produced by SMEs is considerably larger than that by MNCs in Third World countries.

Barring a change in economic and political conditions, FDI in Third World copper production is likely to be largely limited to Chile and PNG over the next five years and only in the case of Chile are new projects in an advanced planning stage. Additional investment is currently being made in the Ok Tedi mine in PNG.[2] There are potential projects in Argentina and Peru in which foreign investors have considerable interest, but economic conditions in Argentina and economic and political conditions in Peru are currently unfavorable for large new projects costing hundreds of millions of dollars. Indonesia's investment climate is satisfactory, but I am not aware of any planned copper projects in that country in which foreign investors have an interest.

The Peruvian SME, Minero Peru, is negotiating joint ventures with several foreign mining companies, but the status of these negotiations is unknown. Minero Peru has plans for developing several large copper deposits, but given Peru's foreign debt position, external financing is unlikely to be available in the foreseeable future. A large project is planned by CVRD in Brazil and this project, along with the CODELCO project mentioned earlier, are the only sizeable Third World SME projects that seem likely to be constructed over the next five years. Both CODELCO and CVRD are probably able to carry out the planned projects from their own resources plus equipment supplier credits, so that central governments external debt problems may not be a barrier.

CAPACITY IN DEVELOPED COUNTRIES

Mine capacity in developed countries is projected to decline between 1985 and 1990, largely due to the permanent closure of most of the nearly 700,000 mt of idle capacity in the U.S. Partly offsetting this reduction in U.S. capacity is a projected increase of 50,000 mt in Australian capacity and the construction of a 65,000 mt capacity mine in Portugal. Little change is projected in Canadian capacity between 1985 and 1996. Some additional capacity may be constructed in Australia and Europe in the early 1990s. However, the share of developed countries in world copper capacity and output is expected to decline significantly.

ADJUSTMENT OF OVERCAPACITY

An estimated 957,000 mt of copper capacity in the market economies was idle in 1985. Of this amount, 684,000 mt was in the U.S., 164,000 mt in Canada, and 105,000 mt in the Philippines (Phelps Dodge, 1985). There were also several countries in which production was well below the rated capacity of mines operating in 1985. The most conspicuous example is Zambia, which had an operating capacity of 610,000 mt, but produced only 475,000 mt. About 125,000 mt of U.S. capacity was permanently shut down in 1985 and some operating capacity was substantially underutilized. Additional U.S. mine capacity totaling 75,000 mt is expected to be added by 1988, mainly by Phelps Dodge. According to presently announced plans, the U.S. would have nearly 1.7 million mt of mine capacity by the end of 1989. Since U.S. mine production was less than 1.1 million mt in 1985 and is expected to decline further by 1990, this would leave at least 600,000 mt idle capacity at the end of 1989. It seems likely that most of this capacity will be permanently idled before 1990. If 500,000 mt of this idle capacity were shut down by the end of 1989, the Phelps Dodge estimate of world capacity (excluding communist countries) at the end of 1989 would be reduced to about 7.9 million mt. This is nearly the same as the 1986 World Bank staff report estimate of world capacity for 1990 (Takeuchi, et.al., 1986, p. 126).

Assuming an annual rate of growth in copper consumption of 1.5 percent per year, how and over what time period is world overcapacity likely to be eliminated? According to the 1986 World Bank staff study, capacity is expected to increase by 400,000 mt between 1985 and 1990. The increases in capacity (mainly in Australia, Chile, Mexico, PNG and Peru) will more than offset the decreases in capacity (mainly in the U.S. and Zambia).[3] If we accept the estimate of 7.9 million mt for world producing capacity (excluding communist countries) for 1990 and assume an annual rate of growth of 1.5 percent in demand for primary (mine) copper over the 1985-1990 period, idle capacity is reduced from 570,000 mt in 1985 (Phelps Dodge estimate) to 232,000 mt in 1990.[4] This is still a substantial amount of idle capacity overhanging the market and, in addition, there would be considerable underutilized capacity. The 1986 World Bank staff study also projects a net increase in world copper producing capacity of about 100,000 mt between 1990 and 1996.[5] Assuming a 1.5 percent annual growth in consumption, by 1996 most of the idle world capacity would have been eliminated and, except for the underutilization of operating capacity, world copper demand and supply would approach equality.

These projections have important implications for the future course of copper prices. Given the existence of considerable excess capacity in 1990, we would expect little increase in prices in constant dollars between 1985 and 1990 and only a modest rise in prices to 1995 or 1996. Thus, the 1986 World Bank staff study projects a price of 74 cents per pound (in constant dollars) in 1995 as contrasted to an LME price of 65 cents per pound in 1986 (Takeuchi, et.al., 1986, p. 128). Beyond 1995 copper prices would need to rise sufficiently to attract investment in new capacity to satisfy the growth in demand to 2000 and beyond. This could mean prices in excess of $1 per pound (in constant dollars) in the late 1990s. A substantial price rise could occur in the early 1990s if the projected 500,000 mt in Chile's copper capacity does not materialize.[6]

The above analysis suggests the following scenario for readjusting the world copper industry. By the early 1990s most of the 700,000 mt of idle U.S. capacity (in 1985) will be permanently closed, although a few mines, including Phelps Dodge's Morenci mine (Arizona) and Kennecott's Bingham Canyon mine (Utah) will be modernized and returned to production. There will also be a limited amount of new U.S. smelting capacity. U.S. mine copper production is projected to decline by about 200,000 mt to 900,000 mt and Canadian production will remain at the 1984 level of 700,000 mt (Takeuchi, et.al., 1986, p. 129). Capacity in Australia and the developing countries is projected to rise, mainly in Chile, Mexico and Peru, partially offset by a decrease in Zambian output. Given my copper consumption growth assumption, by the mid-1990s world capacity and demand should be more or less in balance. At this point there could well be a surge in copper prices in response to a cyclical rise in demand. Prices may need to increase substantially, say, to at least $1 per pound (in constant 1985 dollars) in order to provide the necessary incentive to replace depleted mines and build sufficient capacity to meet the future growth in world demand.[7] Current prices are well below the full marginal cost of expanding capacity.

FUTURE STRUCTURE OF THE INDUSTRY

The world copper industry is undergoing substantial structural change. In the U.S. considerable concentration has been taking place. Anaconda (owned by ARCO) was the third largest U.S. copper producer in the 1970s, but is now out of copper mining. In 1986 Kennecott, formerly the second largest U.S. producer, sold its Chino Mines Company in New Mexico to Phelps Dodge and its Ray Mines Division (Arizona) to ASARCO, retaining only its Bingham Canyon mine. Also in 1986, Newmont Mining, formerly the fourth largest U.S. copper producer, spun off its wholly-owned subsidiary, Magma Copper Company, to Newmont's shareholders, and transferred to Magma its other copper subsidiary, Pinto Valley Copper Corporation (Arizona). These developments, together with several mine closures, have left Phelps Dodge as the dominant copper mining company in the U.S. with about 40 percent of operating capacity. There has also been a substantial decline in oil company participation in U.S. copper mining as evidenced by the withdrawal of Anaconda, the decline in Kennecott's copper output, and the sale of Cyprus Mines by Amoco. The reorganization of the industry has been accompanied by modernization and substantial cost reductions so that it should be able to operate profitably at about 1 million tons per year during the remainder of this century and beyond.

Among the developing country copper producers, Chile has not only become the world's largest producer (excluding the U.S.S.R.), but is projected to account for the bulk of the increase in world capacity over the next decade. The relative decline of production in Zaire and Zambia (both of whose production is entirely state owned) and the projected new foreign private investments in copper mining capacity in Chile, Australia and PNG, signal the reemergence of multinationals in the world copper industry. The bulk of Peru's copper output is currently produced by MNCs and if that country is to develop its rich resources in line with the growth in world demand, it must also depend heavily on new foreign direct investment.

NOTES

1. Information derived from an address by George M. Munroe, Chairman of the Board, Phelps Dodge Corporation, on December 7, 1985 in Tucson, Arizona.

2. There are also two joint ventures involving the government and foreign investors in Turkey--one investor is the German mining firm, Preussag A.S. Metal, and the other is Phelps Dodge. A Japanese company, Panama Minerals Resources Development Ltd., has planned a large copper mine in Panama, but this project is on hold and appears unlikely to be developed in the near future.

3. The increase in capacity of 770,000 mt by 1990 projected by CIPEC is apparently based on announced plans for increases without a realistic appraisal of their likely achievement.

4. This analysis assumes the ratio between primary copper production and total refined production, including secondary, would remain the same over the 1985-1996 period.

5. The 1986 World Bank projections of copper producing capacity for 1990 and 1996 given in the text are "base case" or "most likely" projections. An alternative projection given by the staff study assumes a smaller increase in capacity over the 1985-1996 period with total capacity at the beginning of 1996 of just under 7.7 million mt.

6. For example, most projections of Chile's copper capacity include the construction of the Escondida project (owned by Utah International, RTZ and Mitsubishi) planned to reach 280,000 mtpy by the mid-1990s.

7. A leading copper specialist has suggested to the author that increases in productivity will lower costs sufficiently to permit the necessary expansion in copper capacity without a substantial rise in copper prices in the late 1990s. In view of the sharp decline in costs during the 1980s, he may be right.

REFERENCES

Bureau of Mines (1985) "Copper," Minerals Facts and Problems), Bulletin 675, U.S. Department of Interior, Washington.

Phelps Dodge (1985) Unpublished memorandum of the research staff, New York, November.

Takeuchi, Kenji, John Strongman, Shunichi Maeda and Suan Tan (1986) The World Copper Industry: Its Changing Structure and Future Prospects, Staff Commodity Working Paper No. 15, World Bank, Washington, November.

BIBLIOGRAPHY

Adams, F. G. and S. Klein (eds.) (1978) Stabilizing World Commodity Markets, Lexington Books, Lexington.

American Bureau of Metal Statistics (various issues) Yearbook, New York.

Auty, R. (1985) "Materials Intensity of GDP," Resources Policy, 11, December.

Barsotti, A. F. and R. D. Rosenkranz (1983) "Estimated Costs for the Recovery of Copper from Demonstrated Resources in Market Economy Countries," Natural Resources Forum, 7:2, April.

Bozdogan, K. and R. S. Hartman (1979) "U.S. Demand for Copper: An Introduction to Theoretical and Econometric Analysis," in R. F. Mikesell, The World Copper Industry, Johns Hopkins University Press, Baltimore.

Bureau of Mines (1984) Mineral Commodity Summaries, 1984, U.S. Department of Interior, Washington.

________ (1985a) Mineral Commodity Summaries, 1985.

________ (1985b) "Copper," in Mineral Facts and Problems, Bulletin 675.

________ (1986a) "Copper," preprint from Mineral Facts and Problems.

________ (1986b) Domestic Consumption Trends, 1972-82 and Forecasts to 1993 for Twelve Major Metals, January.

________ (1986c) Mineral Commodity Summaries, 1986.

________ (various issues) Minerals Yearbook.

Carman, J. S. (1985) "The Contribution of Small-Scale Mining to World Mineral Production," Natural Resources Forum, 9, 119-24.

Charles River Associates (1977) The Feasibility of Copper Price Stabilization Using a Buffer Stock and Supply Restrictions from 1953 to 1976, Cambridge.

Commodities Research Unit (1980) "U.S. Copper Production Costs," Copper Studies (monthly publication), New York, June.

________ (1981) "The CRU Long-Term Copper Model," June.

________ (1981) "The Copper Price and Floating Currencies," December.

________ (1982) "Factors Affecting Long-Term Demand," December.

________ (1983) "Production Cost History: Part I," January.

________ (1983) "Production Cost History: Part III," March.

________ (1984) "Varying Patterns of Copper Consumption," April.

________ (1984) "Concentrate Market Roundup," June.

________ (1984) "Secondary Smelters in the U.S.," September.

________ (1984) "1985 Pricing Trends," December.

__________ (1985) "New Mining Projects for the 1990s," October.
Corporacion del Cobre (1975) El Cobre Chileno, Editorial Universitario, Santiago.
Council of Copper Exporting Countries (various issues), Quarterly Review, Paris.
__________ (1985) Survey of Mine, Unrefined and Refined Capacities, Paris.
Crowson, P.F.C. (1982) "Investment and Future Mineral Production," Quarterly Review, CIPEC, Paris, April/June.
__________ (1983) "Aspects of Copper Supply: Past and Future," Quarterly Review, CIPEC, Paris, January/March.
Demler, F. R. (1985) "Copper Market Outlook," Metals, Drexel Burnham Lambert, New York, May.
Evans, B. J. (1986) "How to Assess the Staying Power of World Copper Mines," Engineering and Mining Journal, 187:4, April.
Fischman, L. L. (1980) World Mineral Trends and U.S. Supply Problems, Johns Hopkins University Press, Washington.
Gladwin, T. L. and J. G. Welles (1976) "Environmental Policy and Multinational Corporate Strategy," in I. Walter (ed.), Studies in International Environmental Economics, Wiley, New York, Chapter 8.
Haug, M. (1983) Unpublished address at the Economics Symposium of the American Institute of Mining, Metallurgical and Petroleum Engineers, Washington, November.
Herfindahl, O. C. (1959) Copper Costs and Prices: 1870-1957, Johns Hopkins University Press for Resources for the Future, Baltimore.
Hobson, S. (1984) "The Current and Future Market for Custom Copper Concentrates," Quarterly Review, CIPEC, Paris, October/December.
Intergovernmental Council of Copper Exporting Countries (CIPEC) (1985) Survey of Mine Capacity and Refined Capacities, Paris, June.
__________ (various issues) Quarterly Review.
International Monetary Fund (1985) International Financial Statistics Yearbook 1985, Washington.
Joralemon, I.B. (1973) Copper, Howell-North Books, Berkeley.
Lillie, D. N. and P. A. Abetti (1986) "Technological Opportunities and Threats for Copper," paper presented at Annual Meeting of the Society of Mining Engineers, New Orleans, March 3-6, 1986. Papers published by Society of Mining Engineers, Littleton, Colorado.
Little, A. D. Inc. (1978) Economic Impact of Environmental Regulations on the United States Copper Industry (report prepared for the U.S. Environmental Protection Agency), Boston.

Maizels, A. (1982) Selected Issues in the Negotiation of International Commodity Agreements: An Economic Analysis, UNCTAD, Geneva.

Malenbaum, W. (1977) World Demand for Raw Materials in 1985 and 2000, National Science Foundation, Philadelphia.

Metallgesellschaft (various issues) Metal Statistics Frankfurt.

Mikesell, R. F. (1979) The World Copper Industry, Johns Hopkins University Press for Resources for the Future, Baltimore.

________ (1983) Foreign Investment in Mining Projects, Oelgeschlager, Gunn and Hain, Cambridge.

________ (1984a) "The Evolving Pattern of Contracts and Corporate-Government Relations in Mining with Special Reference to the Asia-Pacific Region," in Research and International Business and Finance, JAI Press, Vol. IV, Part B, pp. 59-95.

________ (1984b) "The Selebi-Phikwe Nickel/Copper Mine in Botswana: Lessons from a Financial Disaster," Natural Resources Forum, 8:3, 279-89.

________ (1984c) Petroleum Company Operations and Agreements in the Developing Countries, Johns Hopkins University Press for Resources for the Future, Baltimore.

________ (1987) Nonfuel Minerals: Foreign Dependence and National Security, University of Michigan Press for the Twentieth Century Fund, Ann Arbor.

Mikesell, R. F. and J. W. Whitney (1987) The World Mining Industry, Allen and Unwin, London.

National Academy of Sciences (1978) Technological Innovation and Forces for Change in the Minerals Industry, Washington.

National Materials Advisory Board (1982a) Mineral Demand Modeling, National Academy Press, Washington.

________ (1982b) Analytical Techniques for Studying Substitution Among Materials, National Academy Press, Washington.

Phelps Dodge (1985) Unpublished memorandum prepared by the Research Staff, New York, November.

Pino, V. (1985) "The Trade and the Reference Price--The Producer's View," Quarterly Review, CIPEC, Paris, July/September.

Pintz, W. S. (1984) Ok Tedi: Evaluation of a Third World Mining Project, Mining Journal Books, London.

Pollio, G. (1983) "The Outlook for Major Metals to the Year 2000: An Updated View," The Journal of Resource Management and Technology, 12:2, April.

Prain, Sir R. (1975) Copper: The Anatomy of an Industry, Mining Journal Books, London.

Radetzki, M. (1979) "The Rising Costs of Base Materials--The Case of Copper," Mining Magazine, April.

________ (1985) "Effects of a Dollar Appreciation on Dollar Prices in International Commodity Markets," Resources Policy, 11:3, September.

________ (1985) State Mineral Enterprises, Resources for the Future, Washington.

Richard, D. (1978) "Dynamic Model of the World Copper Industry," Staff Papers, International Monetary Fund, Washington, 25, 779-833.

Rosenkranz, R. D., E. H. Boyle, Jr., and K. E. Porter (1983) Copper Availability--Market Economy Countries: A Minerals Availability Program Appraisal, BOM Information Circular 8930, U.S. Department of Interior, Washington.

Sacks, H. (1982) "Copper's Metamorphosis from a Merchant's Viewpoint," Quarterly Review, CIPEC, Paris, January/March.

Sassos, M.P. (1986) "Mining Investment 1986," Engineering and Mining Journal, 187:1, January.

Siedenburg, W. G. (1984 and 1985). Copper Quarterly, Smith Barney, Harris Upham, New York, October 19 and June 10.

Sousa, L., A. Weise, and N. J. Themelis (1983) "Assessment of Projected Technological Innovation in the U.S. Copper Industry," Mining Engineering, 35:9 (September, 1271-73) & 10 (October, 1401-05).

Survey of Current Business (1985) "U.S. Direct Investment Abroad," U.S. Department of Commerce, Washington, August.

________ (1986) "Capital Expenditures by Majority-Owned Foreign Affiliates of U.S. Companies," March.

Takeuchi, K., J. Strongman, S. Maeda and S. Tan (1986) The World Copper Industry: Its Changing Structure and Future Prospects, Staff Commodity Working Paper No. 15, World Bank, Washington, November.

United Nations (1974) General Assembly Resolution 3202 (S-VI), New York, May 1.

________ (1980) Mineral Processing in Developing Countries, New York.

U.S. Embassy (1985) "Chile Industrial Outlook Report: Minerals," (mimeo), Santiago, June.

U.S. Federal Trade Commission (1947) "The Copper Industry in the United States and International Copper Cartels," Report on the Copper Industry: Part I, USGPO, Washington.

U.S. International Trade Commission (1984) Unwrought Copper (Report to the President on Investigation No. TA-201-52 under Section 201 of Trade Act of 1974), Washington, July.

Vogely, W. S. (ed.) (1975) Minerals Materials Modeling: A State of the Art Review, Resources for the Future, Washington.

Wagenhals, G. (1985) "Econometric Copper Market Models," Quarterly Review, CIPEC, Paris, January/March.

Walde, T. (1985) "Third World Mineral Development and Crisis," Journal of World Trade Law, 19:1, January/February.

Walter, I. (1975) International Economics of Pollution, Macmillan, London.

Weise, A., N. J. Themelis and N. E. Guernsey (1983) Technological Innovation in the Copper Industry, (BOM), U.S. Department of Interior, Washington, March.

World Bank (1983) The Outlook for Primary Commodities, Staff Commodity Working Paper No. 9, Washington.

_____ (1984) Prospects for Major Primary Commodities: Vol. IV, Washington, September.

_____ (1985) Commodity Trade and Price Trends, Johns Hopkins University Press, Baltimore.

_____ (1986) "Half-Yearly Revision of Commodity Price Forecasts and Quarterly Review of Commodity Markets for December 1985," (mimeo), Washington.

World Mining Equipment (1986) "Modernization at Utah," 10:2, February.